# Changing Digital Geographies

Jessica McLean

# Changing Digital Geographies

## Technologies, Environments and People

palgrave
macmillan

Jessica McLean
Department of Geography and Planning
Macquarie University
Sydney, NSW, Australia

ISBN 978-3-030-28306-3 ISBN 978-3-030-28307-0 (eBook)
https://doi.org/10.1007/978-3-030-28307-0



Cover credit: Photo by Katie McLean

This Palgrave Macmillan imprint is published by the registered company Springer Nature Switzerland AG
The registered company address is: Gewerbestrasse 11, 6330 Cham, Switzerland

# Acknowledgements

Thank you to Rachael Ballard and Joanna McNeil at Palgrave Macmillan for their excellent editorial support.

Thank you to Macquarie University for employing me during the writing of this book.

Thank you to the research participants who so generously shared their knowledge, time and energy during interviews for this book.

Thank you to the Critical Development and Indigenous Geographies research cluster, Department of Geography and Planning Macquarie University, for reading, commenting on and critiquing the chapter on decolonising digital geographies.

Thank you to Dr. Linda Steele, University of Technology Sydney, for reading, critiquing and sharing useful research materials, for the chapter on people with disabilities and digital geographies.

Thank you to Dr. Sophia Maalsen, University of Sydney, for crucial early conversations about digital geographies and for walking along winding academic roads with me, real and more-than-real.

Thank you to Alan Vaarwerk at Kill your Darlings for publishing my essay *Those Anthropocene Feelings* that helped to extend my thinking for Chapter 8 *Feeling the Digital Anthropocene.*

Thank you to my family for helping out with Lorenzo and my academic world overall—Mum, Dad, Rachel, Sharon, Daniel, Guida, Lizzy, Dan B., Gabe, Josh, Katie, Nomie, Han, Josh B., Joe, Sophie, Eloise, Jacinta, Dominic, Abel, Aidan, Mae, Josie and Anna. And to Kelly Yates, Abbie Hartley, Phoebe Bailey, Liz Starr and Linda Martin for same.

Big thanks to Katie McLean for generously taking the cover image and making it fit for purpose.

Thanks to Rohan Mackenzie for careful reviewing and editing of Chapter 1, and for patient conversations during the making of this book.

And this book is for Lorenzo, my son, who has been challenging and changing my digital geographies for a few years now, in the most surprising and excellent ways.

# Contents

# List of Figures

# List of Tables

# 1
# Introduction

## Changing Digital Geographies

Digital geographies are constantly changing as individuals, community groups and institutions take advantage of these amorphous contexts. Users of digital technologies are subject to, and participants in creating, innumerable digital changes, while corporations continue to render devices and software obsolete to increase profits, and governments are slowly becoming involved in regulating the digital. Digital technologies are providing new opportunities for communication and connection, while simultaneously deepening problems associated with isolation, global inequity and environmental harm, contributing to shifting digital geographies.

The work of those trying to achieve digital, environmental and social justice also refigures the digital. For instance, GetUp! in Australia is 'an independent movement of more than a million people working to build a progressive Australia and bring participation back into our democracy' that has used digital technologies to produce political changes. In a conversation with Lyn Goldsworthy, GetUp! Board Member, I learnt about


J. McLean, *Changing Digital Geographies*,
https://doi.org/10.1007/978-3-030-28307-0_1

how digital work is at the core of their interventions and I mentioned the growing membership of this organisation:

> Yeah, well, and is that not surprising? I mean, it is a citizen-led organization. The GetUp! executive does not decide in a vacuum what it should work on. I mean, I found it quite difficult initially because I come from a "long-term strategic, pick up an issue and then follow it through until you win" approach. And it took me a little while to realize that GetUp!'s role is to galvanize issues and galvanize the public to become involved in those issues. And then if someone else picks them up to finish them off, fine. But our job is more the one-year, two-year, three-year work… not saying quick fix because that's not the right word, but it's that initial get-everybody-engaged work. (Lyn Goldsworthy interview, November 2018)

GetUp!'s work is funded by donations from citizens, and spans campaigns as diverse as bringing refugees to Australia, to pushing for solar power to become the bulk of energy consumption. The progressive politics that GetUp! advocates has attracted repeated political attacks, including the creation of a counter-activist group called 'Advance Australia' in 2018, founded by conservative individuals who wanted to curtail GetUp!'s reach. In the lead up to the Federal election in 2019, Advance Australia created an inflatable mascot called 'Captain GetUp!' with a Twitter account that was meant to satirise GetUp!'s approach. The Twitter presence of Captain GetUp! states that it 'has arrived to tell you what to think and how to act this election!' With fewer than two thousand followers, Captain GetUp! has not achieved significant reach, except as an object of ridicule itself; a parody account for Captain Getup! has more followers than the original. The parody of the satire out-performed it, and Getup! continues to grow as an organisation.

The polar possibilities that shape digital geographies—of generative and destructive processes in digital spaces—produce a space that could be understood as 'more-than-real'. Here, I am using the 'more-than-real' idea as a political strategy and to examine how digital technologies, humans and environments interact to produce changing digital geographies. As a political strategy, the more-than-real idea carries the potential to build on arguments on the materiality of the digital and the agency of non-human digital actors. In 1987, Brian Massumi, speaking to Deleuze and Guattari in an article on the simulacrum, touched on the idea of the 'more-than-real':

> The reality of the model is a question that needs to be dealt with…The alternative is a false one because simulation is a process that produces the real, or, more precisely, more real (a more-than-real) on the basis of the real. (Massumi 1987, 93)

Massumi is arguing that there is no real and simulation that exist in opposition to, or distinct from, each other, and that we make, and remake, the real from already 'real' things. In effect, a simulation, or representation, is as real as the original 'real'—except that there never is an original. At the time of Massumi's writing on the realer than real, digital geographies were only emerging but the patterns of thinking about reality and ways of being that he stipulates resonate with this space.

The 'more-than-real' is assembled from multiple elements and produces powerful effects; as a concept, the more-than-real is inspired by geographers' work on the more-than-human (for example see Whatmore 2002). More-than-human work emphasises material interconnections of humans and non-humans in the world, and challenges the binary thinking that dominates western thought. Similarly, the more-than-real concept inverts the diminishing that accompanies use of the terms 'virtual' and 'immaterial' as applied to digital spaces, moving away from tendencies to place these realms as inferior and subordinate to the 'real'. Drawing on Eve Kosofsky Sedgwick's writing in *Touching Feeling* (2003), the more-than-real sits *besides* thinking about the 'real' and digital geographies, not beyond or behind such work, in an effort to think through dilemmas of digital spaces in a nondualistic way. Following Sedgwick, rather than reproducing binaries and resisting nondualistic thought, the more-than-real sits with ideas about how digital geographies are made, and are making us. The more-than-real concept does political, cultural, social and environmental work in assembling facets that cut across space/time compressions to produce polar activity—excesses of productive and destructive forces of social change with material entanglements (McLean 2016). The more-than-real, then, is also building on new materialist thinking on the political ecology of things, including Jane Bennett's (2009) landmark 'Vibrant Matter' and conceptualisations of digital spaces as forming a sort of public sphere (Papacharissi 2002).

In thinking through changing digital geographies, I offer an analysis of digital spaces as more-than-real, rather than unreal, recognising the affective and emotional forces that co-produce the digital, contributing to the growing digital geographic literature. Ash et al. (2019) co-edited a wide-reaching volume on 'Digital Geographies' and argued that there is value in drawing on previous thinking about shifts in geographic work to inform the current wave of digital studies. Reminiscent of thinking relating to the earlier cultural turn in geography, Ash et al. (2019, 5) suggest that it is useful to think of digital geographies 'as a turn *towards* the digital as object and subject of inquiry in geography, and as a simultaneous inflection of geographical scholarship *by* digital phenomena'. Conceptualising a digital turn, rather than arguing for a new sub-discipline within geography, suggests that digital geographies are an inflection, rather than a transformation, of geographic thought and practice. I argue that we do not yet know the extent of the shifts that digital geographies may offer, as important questions are being formed about how digital geographies work, and how societies and environments are remaking these spaces. For instance, do we consider emotion, affect and ontologies as central to this digital turn? And how are we remaking the digital through social movements and cultural practices, that in digital spaces combine the personal with the political, the everyday with concerns relating to digital justice? We can continue to interrogate what digital geographies mean for navigating and producing global environmental changes, and perhaps whether we can decentre the human in understanding digital rights in this research area. Building on earlier work by Pickerill (2003) and Kitchin and Dodge (2011), this book aims to contribute to this growing and timely conversation.

Latour analyses human and technology relations in 'Love your monsters: Why we must care for our technologies as we do our children' and offers the narrative of Frankenstein's misadventure with his creation as a metaphor for how humans engage with technology. Latour (2014) exhorts humans to take better care of technologies and avoid repetition of Frankenstein's mistaken abandonment of his charge. Similarly, this metaphor can be applied to our digital technologies, inviting corporations, governments and individuals to not neglect the unwieldy and powerful technologies that co-produce our digital lives. The technologies that comprise the digital are made of 'entanglements of all those things that were once imagined to be separable—science, morality, religion, law, technology, finance and politics' (Latour 2014, no page numbers). These entanglements are, in some ways, making monsters of humans, the digital and non-humans, of multiple sorts. It is appropriate to think about the ways that digital monsters emerge in the more-than-real, and to delineate the who, what, where, why and how of these digital geographies, however nebulous, that form them.

Some standout digital monsters are framed as the 'frightful five' by Manjoo (2017) and are collectively worth trillions: Amazon, Apple, Alphabet (owner of Google), Facebook and Microsoft. These particular digital monsters are not subject to strong government regulations and are deeply entrenched in aspects of everyday lives, in most parts of the Global North and increasingly so in the Global South. The limitations of the digital are visible in the exacerbation of uneven geographies of representation and participation in the digital (Graham et al. 2015). The Global North continues to dominate digital information technologies in terms of making, and being the subject of, most digital data. Ballatore et al. (2017) identify digital hegemonies in representations of place with respect to the Global South, signified by less locally produced digital content than the North in Google searches. The North is finding new ways to continue exploiting the South, but the South(s) are also finding new ways to assert agency (Milan and Treré 2019) and countering neocolonial digital geographies is happening on multiple fronts. Further digital dilemmas emerge if we fetishise big data sets that can potentially seduce researchers and policy-makers alike with the scale and scope of data to analyse (for example, see big data analysis by Stephens-Davidowitz 2017). Big data is a seductive pool of information for analysis but Kwan (2016) urges that geographers use caution with these sources of information and what they mean, and for us to consider the impact of algorithms that drive aspects of big data generation. Similarly, Milan and Treré (2019, 328) argue that we must understand Big Data from the South and that this 'entails the engagement with a plurality of uncharted ways of actively (re)imagining processes of data production, processing, and appropriation'. Attempting to decentre the Global North is a key part of this transformation.

The most well-known damaging aspects of the digital might be the troll—that digital creature which emerges at particular spaces and times, to fight disparate and sometimes organised campaigns, in groups such as 4chan or Anonymous (Coleman 2014). These trickster characters, similar to hackers (Nikitina 2012), are slippery aspects of digital spaces. Coleman (2014) ethnographically followed the work of Anonymous and found that her insider–outsider status in relation to the group became part of the narrative of who and what Anonymous is. Rather than being a one-dimensional digital deviant, Anonymous works in tricky ethical spaces according to Coleman (2014), some damaging, others not so. We could point to Donald Trump's Twitter use as a monstrous spectacle relying on incivility and hyperbole (Lee and Xu 2018). Other digital monsters might include the Australian government—keeping metadata for two years after it has been created and having massive digital failures with the census and datafication of social payments (Galloway 2017).

The social media presence that corporations enable and individuals cultivate can be monstrous in their addictive qualities built from classical conditioning, while our employers can also be framed as introducing digital dilemmas with their reliance on the tentacles of the digital, extending into private domains and outside of formal work hours, producing troublesome 'intimate geographies of the digital' (Richardson 2016, 14). I have briefly contributed to conceptualisations of the limitations of the digital by offering versions of digital monsters here, so that we can reflect upon, and think of, ways that the digital works, and to highlight the breadth and depth of troublesome digital ways of being. The more-than-real can produce polarised and contradictory relations that are at least partly shaped by emotion and affect in human–technology relations. After all, as Ahmed (2013, 18) attests 'Emotions are shaped by contact with objects'. It is also important to think of ourselves, and our desire (and need?) to be digitally engaged, wherever and whenever, as monstrous habits—as tenuous and troubling as this 'our' category might be.

But it also matters that to *do something* digitally can involve little physical or externalised action: the critiques of armchair activism, or slacktivism (Goldsborough 2011), have struck on something here—although they have not taken their argument to the point of full explanation for *why* some digital actions are sometimes so effective, even if they seemingly do not require significant effort. To sign a petition, like something on Facebook, or post a tweet using a particular hashtag, does not require the same level of deliberate exertion as joining a protest on a public street, blockading a farm property to stop intrusive gas exploration, or becoming part of a picket line. And it is that relative ease, of activism through the digital, that enables even the smallest surges of emotion and affect to generate something—those digital gestures—and then cumulatively, individual minor acts coalesce to form moments, and possibly movements, that stem from, but are not limited to, the more-than-real.

Returning to the story of Frankenstein and the monster, Latour sees Frankenstein's problem as arising from a lack of care for that which he created and failed to attend to:

> Frankenstein lives on in the popular imagination as a cautionary tale against technology. We use the monster as an all-purpose modifier to denote technological crimes against nature. When we fear genetically modified foods we call them "frankenfoods" and "frankenfish." It is telling that even as we warn against such hybrids, we confuse the monster with its creator. We now mostly refer to Dr. Frankenstein's monster as Frankenstein. And just as we have forgotten that Frankenstein was the man, not the monster, we have also forgotten Frankenstein's real sin.
>
> Dr. Frankenstein's crime was not that he invented a creature through some combination of hubris and high technology, but rather that he *abandoned the creature to itself.* (Latour 2014, paragraphs 3–4, italics in original)

Analogously, if humans fail to ethically engage with the digital—considering questions of justice and sustainability as we expand our digital lives—the more-than-real can potentially slip further away, out of any semblance of control. The more-than-real concept comes from concern about the unfurling of digital geographies without check. The language we use in everyday life for the digital invokes its immateriality: including terms like the virtual and the opposite of 'IRL' (in real life). If the digital is not framed as a version of the real, then attempting to negotiate and control it, to manage its unwieldy possibilities, is next to impossible.

Tendencies to un-real digital spaces are evident in commentary on the power, or otherwise, of digital activism. For instance, Gladwell (2010) wrote an oft-cited essay on the paucity of online activism and confused social media interactions as 'not real'. He wrote that 'The evangelists of social media…seem to believe that a **Facebook friend is the same as a real friend** and that signing up for a donor registry in Silicon Valley today is activism in the same sense as sitting at a segregated lunch counter in Greensboro in 1960' (Gladwell 2010, paragraph 19, emphasis added). Gladwell goes on to say that Facebook reinforces people's sense of having contributed to social change by pursuing superficial change. He argues that digital activism motivates people to 'do the things that people do when they are not motivated enough to make a real sacrifice'. Again, the binary demarcation of the digital as not real and ineffective is produced in the term 'real sacrifice'. Digital spaces are paradoxical and networked and yet binaries emerge at certain times when thinking about, and talking of, digital engagements and entanglements. Or, as Karpf (2016) suggests, it is tempting to either celebrate the glorious potentials of digital activism or bemoan rampant clictivism, as Gladwell does above. Another example of not-realing appears in Pesce's (2017) essay on the way Facebook is morphing everyday lives when he argues that 'The real world is about to disappear. It all begins with fake news'. Pesce is stating that we are on a precipice of reality falling away, with all that is solid melting into air, as Marx and then Berman assuaged, and that fake news is the beginning of this. Fake news is, of course, a product of the digital. Still, this dystopic forecast assumes that a singular 'real' ever existed, something Deleuze and Guattari, and Massumi, contest, and places the digital as an all-powerful agent of change rather than as technologies that are already being reckoned with.

I offer the more-than-real in a similar way to how the more-than-human has offered a new language for relational analysis of human–nature connections. Further, I depict digital spaces as more-than-real, rather than unreal, to elevate recognition of the affective and emotional forces that co-produce the digital. The desire to do something in digital spaces produces social, cultural, economic and environmental changes that are real, and can challenge normative spatial relations, sometimes in surprising ways, as this book will show. The not-realing of digital spaces can lead to a misunderstanding of what is happening there as the amplification of emotion and affect forms ad hoc publics (Bruns and Burgess 2011) or more persistent issue publics (Kim 2009). The role of emotion and affect in the digital is, at first glance, confusing and the more-than-real may provide another tool for working with these messy, complicated terrains.

Frequently, corporations, governments and individuals tend to minimise the impacts of digital spaces when thinking about global environmental changes since we conceive of the digital as intangible and not substantive. My work joins the critical geographic thinking that is already happening in this space; for example, Büscher (2016) gives insights into the way nature 2.0 is conceptualised and engaged with in conservation action. He defines nature 2.0 as 'co-creative' technologies with two-way relations where information is produced, consumed and communicated at the same time, such as crowdfunding to produce a conservation result like more protected areas. Nature 2.0 has the potential to increase democratic processes in society through enabling people to take control and initiate social, environmental and cultural changes. Büscher (2016) gives the case of the 'elephant corridor' where online hopes for supporting and creating a new space to protect elephant habitat resulted in the crowdfunding 'of €430,000 for the establishment of an elephant conservation and migration corridor from Chobe National Park in Botswana via the Caprivi Strip in Namibia to the Kafue flats in Zambia' (Büscher 2016, 164). However, this did not translate into an effective conservation outcome for elephants and Büscher concludes that in nature 2.0 'it is harder to see the disjunctures and hierarchies but also to whether there is **any real incentive** to try to see and understand these' (Büscher 2016, 173, emphasis added). It definitely is harder to see disjunctures and hierarchies in nature 2.0, yet depicting social media efforts such as crowdfunding as devoid of 'real incentives' evokes the way we sometimes position the digital as not real, reproducing binary ways of thinking about online actions as immaterial and with less accountability than face-to-face or offline modes.

If we consider emotion and affect and how these facets intermingle with action in and on digital spaces, different geographies of responsibility (Massey 2004) emerge. Massey (2004) asked us to think about space relationally and, similarly, a relational analysis of digital geographies must account for how emotion and affect work therein. Affect, according to Shouse (2005, xx), is a 'non-conscious experience of intensity' while emotion is defined as an outward expression of a feeling. Pile (2010, 9) reads how affect and emotion are placed in geographic work and usefully points to how 'Like emotions, affects matter—but they cannot be grasped, made known or represented'. This intangibility coincides with the sometimes-perceived immateriality of digital spaces, however, it is now well established that the latter is a conceptual mistake. We use smartphones, tablets, and computers to access and make the digital, which has a materiality that has been widely acknowledged (Pink et al. 2016). Further, the environmental and social costs of the inbuilt obsolescence of these technologies are important, although often ignored, aspects of the more-than-real. Framing these as more-than-real can challenge the elusive qualities of the interplay of affect, emotion and the digital. As a political strategy, the more-than-real concept offers a way to consider interactions of different forces to produce change—such as political interventions that involve digital tools and the transformation of emotion into affect.

By emphasising *changing* digital geographies, this book considers the way technologies are situated in broader social, cultural, economic, environmental and political contexts. Multiple forces change digital geographies and this book focuses on some, but not all of those, including digital justice movements, environmentalism and feminism. At the same time, geographies of digital spaces can change us, the environment, and how we understand human–nature relations. I am examining the troubling *and* satisfying ways that digital geographies emerge while at the same time avoiding characterisations of technology as utopic or dystopic. Technological determinism has attracted critique for positioning technologies as drivers of social change (Gunkel 2003). This book builds on these critiques to position geographies of digital spaces in relation to shifting social, environmental, cultural and economic dynamics. Digital geographies do not determine social and cultural forces, as they are a part of the broader picture of how transformations can come to pass.

The more-than-real draws out specific aspects of human, environment and cultural relations that are sometimes overlooked, and forwards an intersectional approach. Given that I am looking at fluid, hybrid institutions and spaces (Foth et al. 2013), it is important to take into consideration how affect and emotion co-produce ways of being 'digital' as well as more traditional political processes. This book examines the relationships between humans, digital technologies and environments to bring together discourses and materialities of digital geographies that are sometimes left separate. In doing so, I aim to join conversations on how and why digital geographies are changing. The use of a digital geographic lens in this context extends a grounded analysis of social and cultural processes in particular ways: rather than separating social injustices from environmental dilemmas, these are considered in relation to each other. Digital dilemmas are visible in such an integrated approach, contributing to the growing interdisciplinary literature on the changing nature of human–technology relations, including with respect to gender (Parry et al. 2019).

## Methodology

The research that shapes this book involved bricolage methods including semi-structured interviews, participant observation, content and discourse analysis of social media, and analysis of digital archival material. The participant observation research—at conferences, in digital spaces, and at public events—has been useful in terms of seeing how debates in digital technologies are playing out and gaining invaluable insights on corporate technological companies and their modus operandi. For example, the Human Rights and Technology Conference in Sydney during 2018 (hosted by the Australian Human Rights Commission) included representatives from Microsoft, YouTube and ex-Mozilla employees, as well as Australian government representatives, digital activists, academics and interested members of the public. The major fault lines in digital geographies were apparent at this event as issues of justice, freedom, privacy and equity played out in broad discussions about digital technologies, including on Artificial Intelligence (AI) and technologies for people with disabilities.

Participant observation research also enabled meeting people to interview for this book. In-depth conversations were held during 2018 and 2019 with activists, bureaucrats and academics who are working to change digital spaces. I met the leader of the Digital Gap initiative, a not-for-profit organisation that is campaigning for universal access in digital technologies, at the Human Rights and Technology Conference and was able to interview her at a later time. Fifteen generous research participants (listed in the Appendix) shared their knowledge on how digital spaces are transforming and what further changes are required to enable safe, equitable and sustainable digital geographies to come to the fore. This qualitative information was coded for key themes and has provided a rich tapestry of lived experience, professional know-how and research-informed data to shape this book.

Content and discourse analysis of social media was used to analyse digital dialogue around key events relating to this research and to gain insights into general patterns of cultural and social processes in digital spaces. I do not use big data or extensive quantitative analysis of trends in social media: rather, the nuance of context-based analysis of digital discourse and selective searching of key terms during particular time periods was used for this research (following Kelsey and Bennett 2014; Pedersen and Lupton 2018). The research aims of this book intend to emphasise the 'reflexive and creative agency of human actants' (Rose 2016, 766), *as well as* non-human actants, to bring these forces into dialogue. As Elwood (2010) stipulates when talking about the geoweb, digital technologies do no work in the same way as earlier information sharing modes. The possibilities that digital technologies afford in 'opening up' previously closed production processes further justify the reflexive qualitative research approach used in this book.

A quick note on terminology. I use 'digital geographies' interchangeably with 'digital spaces' in this book. Digital spaces are more than computer code—the term is inclusive of technologies, methods, performances, communication tools, and practices that enable being digital. Digital geographies, therefore, are not neutral as they retain the ideologies, politics and practices that comprise its constitutive parts, yet likewise the digital is also appropriated to resist, disrupt and parody, and play a part in activism and intervention (McLean and Maalsen 2019). Digital geographies are uneven, multiply produced and partial. 'Digital technologies' refers to any technology using digital instructions, including smartphones, robots, computers, and encompasses software and hardware for these devices.

## Book Themes and Chapter Summary

The book begins by positioning the theoretical framework relating to the more-than-real in Chapter 2, integrating literature from cultural geographic, media and communication and gender studies that speak to digital geographies. The agency of humans in remaking digital spaces is a focus here but also of more-than-humans that demand a different way of approaching the digital. Digital geographies bring together multiple scales, from the micro to the macro, and produce new spectacles such as the Anthropocene. The more-than-real framing helps to see how massive offerings such as this new epoch are a part of ongoing reworkings of human and nature relations, and that just as they have been assembled, they can also be taken apart and remade.

Chapter 3 analyses my participation in the 'Human Rights and Technology' (2018) conference and shows how digital rights are being renegotiated by governments, civil society and corporations. Digital technologies were frequently reduced to AI at the conference—a slippage that reflects a dominant focus in the tech industry. The liberatory and restrictive aspects of digital technologies were a key theme throughout, as participants described changing digital geographies.

The multiple environmental implications of increasing reliance on digital technologies were largely left out of the discussions.

Chapter 4 outlines arguments about digital rights and shows how corporations, governments, civil society and academia all play roles in how these digital rights are understood and challenged. The more-than-real is not fixed or static—it is comprised of possibilities and heterogeneities, akin to other spatial imaginings such as those offered by Massey in her manifesto *for space*. Chapter 4 also offers a perspective on digital justice that shows how more-than-real geographies emerge from the trivial and mundane to produce substantive interventions. Stories of digital justice indicate how a groundswell of ethical action is trying to change digital geographies.

Chapter 5 outlines how digital spaces are facilitating new avenues for Indigenous peoples' resistance to colonial hegemonies and possibilities for community building. But it also traces how Indigenous people experience intense levels of racism and violence simply for being Indigenous in the digital. Creative campaigns facilitate political activism in social media and the challenging of colonial power—including moments such as #Indigenousdads. The hard edges of the more-than-real are sharply felt by Indigenous peoples but they are also resisted and remade.

Environmental action in Australia, and specifically mostly related to climate change, is the focus of Chapter 6, from the crowdfunding by everyday people that created the Climate Council to the digital tools that enables visibility of other environmental NGOs. The emotional landscapes in digital spaces that inspire these movements include anger, outrage, grief and transformations to hope. The more-than-real is being used and remade for environmental work that attempts to mitigate the Anthropocene, as fraught as that work may be.

Chapter 7 canvasses how digital geographies are sculpted by digital technologies that carry assumptions about their purported efficiency and sustainability. Discourses of 'green IT' have circulated for decades, meanwhile our consumption of digital devices is increasing and frequently disconnected from accountability of environmental implications of such consumption. The more-than-real amplifies the dissonances of capitalist societies as the materiality of digital geographies is contingent and easily ignored.

Chapter 8 brings together perspectives on emotion and affect with the (digital) Anthropocene. The digital is as contingent as the Anthropocene, I argue, and both are networked, material and abstracted spaces and concepts. We could read both the digital and the Anthropocene as more-than-real as they work with and produce challenging geographic relations and ways of knowing and being that blend rational, emotional and affectual action.

The contradictory possibilities of the more-than-real are present at a global scale with the expansion of access to digital spaces and the deepening reach of the digital in everyday life. Chapter 9 gives an overview of how feminist interventions, from #MeToo to #everydayfeminism, are persistently amplifying sexism and misogyny and reforming digital geographies in diverse contexts and with different opportunities. From China to Turkey, Australia to India, feminist digital activism is taking advantage of the more-than-real as another way to achieve better conditions for marginalised peoples and, at the same time, developing intersectionality.

Chapter 10 expands on digital feminisms in Australia and shows how an attunement to emotion and affect continue to propel this work. The intermingling of logical arguments to stop sexism and misogyny with emotive appeals for action underlies some explicit feminist interventions in the digital. This is a seemingly never-ending task as digital technologies include affordances for relative safety and new opportunities for abuse or violence. Further, the environmental framings of digital geographies by those working on and with feminist digital spaces adds an inflection that integrates important qualities of the more-than-real.

The last case study chapter looks at digital gaps in the more-than-real by sharing insights from disability activists on the politics of inclusion and exclusion. The rapid ascent of digital technologies frequently does not include attainment of universal access, extending serious dilemmas for people with disabilities. People with disabilities can be marginalised and/or gain greater independence from different technological interventions. Drawing on a biopolitical critique of neoliberal inclusionism, no amount of universal design will enable access for all people with disabilities if social, political, legal and environmental structures are not reformed to prevent exclusions.

To conclude, I draw together insights from the cases in preceding chapters to delineate the multiple ways in which the digital is made, and remade, as more-than-real. This book offers a partial perspective on the ways in which technologies, environments and people are affected by, and co-produce, digital geographies that suggests we are already building more ethical digital spaces. Wide-reaching structural changes are urgently needed: corporations and governments need to refigure digital geographies in socially and environmentally just ways.

## More-Than-Realing

The cover of this book shows a hand holding a smartphone, capturing an image of a flower, taking it out of context from its vegetation. The flower is held static and also as representative of an environment that it may or may not belong to. There is only one flower in the image but in the smartphone image that will be formed, the flower dominates and denotes the entire scene. The photograph of the smartphone taking the image reminds the viewer that the flower, a callistemon or bottlebrush bloom, is being held out, distinct to and yet connected with the nature that it is embedded within by a digital technology. The flower is *besides* the foliage that makes it possible, as Sedgwick might say, in this digital geographic rendering. The more-than-real reframes that which already exists—non-humans and humans alike—and amplifies, disconnects, and distorts in that reframing. The flower is not all there is, and the green foliage that enables it is partly captured in the smartphone reimagining; the callistemon becomes more-than-real in this moment of human–technology engagement.

It is no longer feasible to take for granted the more-than-real and its environmental, social, cultural, economic and political implications. The Anthropocene moment that has emerged from Global North excesses shows how global environmental change is now partly driven by humans (Steffen et al. 2007). Rather than accept that this is how and where we all must dwell, there are changes that we can make (Haraway 2015), including appreciating the breadth and depth of digital geographies, and their environmental and social costs and benefits.

The more-than-real concept can help mobilise such work as it draws attention to the minimisation of impacts and potentialities of the digital. The stickiness of emotion in digital geographies is clear in these manoeuvres.

I have never been an early-adopter, and I am not highly technologically skilled. My son plays digital games proficiently and I see him watching Netflix series with delight, following YouTubers closely and enjoying eliminations in Fortnite with his friends. We manage this digital space in tension: he wants more all the time, I'm constantly calling for shorter and less frequent spells in these more-than-real worlds. But then I'll also happily binge watch a series for a couple of hours so I can't pretend my own digital connections are any better. I grew up as a white person in Mudgee, then a town of seven thousand people in rural southeastern Australia, a place in settler colonial Australia with a violent, and often silenced, invasion history. Wiradjuri people and Country live on, in and around Mudgee, reinvigorating practices including cultural burning to make their Country healthy again. I was privileged in that my parents were teachers and we lived in a place with prosperity and stability; along with my nine siblings, we had television and videos and occasional access to computers. My Dad would bring his class computer home from school during school holidays and, on rainy winter days, we'd play Where in the World is Carmen Sandiego? I was sixteen before I used the World Wide Web for the first time, a few years before the turn of the twenty-first century. Since then, I've engaged with digital technologies as required rather than as someone seeking out the latest and best of hardware and software and do not identify as a technologist. So I bring an ambivalent perspective to this book on changing digital geographies: I am concerned about what we are doing with these spaces and enjoy their affordances at the same time; I think we can do wonderful things in digital geographies and see that they are contributing to global environmental changes that are far from desirable; I want to be a part of curtailing corrosive human–digital relations and amplifying the generative.

There are gaps and absences in this book, stories not told, arguments missed; this research is partial. I was unable to talk with corporations that are deeply involved in making our digital geographies and many people I approached for interviews were not available. Limitations of this book include those shaped by my positioning, perspective and capacity. I enjoyed researching and writing it, however, and hope it may do some of the work it sets out to do. I continue to be inspired by Massey's (2005, 12) thinking about geographies as always in the process of being made: 'This is a space of loose ends and missing links. For the future to be open, space must be open too'.

## References

Ahmed, S. (2013). *The Cultural Politics of Emotion*. Edinburgh: Edinburgh University Press.

Ash, J., Kitchin, R., & Leszczynski, A. (2019). *Digital Geographies*. London: Sage.

Ballatore, A., Graham, M., & Sen, S. (2017). Digital Hegemonies: The Localness of Search Engine Results. *Annals of the American Association of Geographers, 107*(5), 1194–1215. https://doi.org/10.1080/24694452.2017.1308240.

Bennett, J. (2009). *Vibrant Matter: A Political Ecology of Things*. Durham, NC: Duke University Press.

Bruns, A., & Burgess, J. E. (2011). The Use of Twitter Hashtags in the Formation of Ad Hoc Publics. In *Proceedings of the 6th European Consortium for Political Research (ECPR) General Conference 2011*. Retrieved June 24, 2019, from https://eprints.qut.edu.au/46515/.

Büscher, B. (2016). Nature 2.0: Exploring and Theorizing the Links Between New Media and Nature Conservation. *New Media & Society, 18*(5), 726–743.

Coleman, G. (2014). *Hacker, Hoaxer, Whistleblower, Spy: The Many Faces of Anonymous*. London: Verso.

Elwood, S. (2010). Geographic Information Science: Emerging Research on the Societal Implications of the Geospatial Web. *Progress in Human Geography, 34*(3), 349–357. https://doi.org/10.1177/0309132509340711.

Foth, M., Parra Agudelo, L., & Palleis, R. (2013). Digital Soapboxes: Towards an Interaction Design Agenda for Situated Civic Innovation. In *Proceedings of the 2013 ACM Conference on Pervasive and Ubiquitous Computing Adjunct Publication* (pp. 725–728). ACM. Retrieved June 10, 2019, from https://www.uni-siegen.de/infme/start_ifm/veranstaltungen/marcus_foth_-_position_paper.pdf.

Galloway, K. (2017). Big Data: A Case Study of Disruption and Government Power. *Alternative Law Journal, 42*(2), 89–95. https://doi.org/10.1177/1037969X17710612.

Gladwell, M. (2010). Small Change: The Revolution Will Not Be Tweeted. *The New Yorker*. Retrieved December 4, 2018. https://www.newyorker.com/magazine/2010/10/04/small-change-malcolm-gladwell.

Goldsborough, R. (2011, January 10). 'Slacktivism' Is Becoming the New Activism: Community College Week, 13. *General OneFile*. Retrieved February 13, 2019, from http://link.galegroup.com/apps/doc/A248188254/ITOF?u=macquarie&sid=ITOF&xid=331d1a41.

Graham, M., De Sabbata, S., & Zook, M. A. (2015). Towards a Study of Information Geographies: (Im)Mutable Augmentations and a Mapping of the Geographies of Information. *Geo: Geography and Environment, 2*(1), 88–105.

Gunkel, D. (2003). Second Thoughts: Toward a Critique of the Digital Divide. *New Media & Society, 5*(4), 499–522.

Haraway, D. (2015). Anthropocene, Capitalocene, Plantationocene, Chthulucene: Making Kin. *Environmental Humanities, 6*(1), 159–165.

Karpf, D. (2016). *Analytic Activism: Digital Listening and the New Political Strategy*. London: Oxford University Press.

Kelsey, D., & Bennett, L. (2014). Discipline and Resistance on Social Media: Discourse, Power and Context in the Paul Chambers 'Twitter Joke Trial'. *Discourse, Context and Media, 3*, 37–45.

Kim, Y. M. (2009). Issue Publics in the New Information Environment: Selectivity, Domain Specificity, and Extremity. *Communication Research, 36*(2), 254–284.

Kitchin, R., & Dodge, M. (2011). *Code/Space: Software and Everyday Life*. Cambridge, MA: The MIT Press.

Kwan, M. P. (2016). Algorithmic Geographies: Big Data, Algorithmic Uncertainty, and the Production of Geographic Knowledge. *Annals of the American Association of Geographers, 106*(2), 274–282.

Latour, B. (2014). Love Your Monsters. *Next Nature*. Accessed December 4, 2018, from http://www.nextnature.net/2014/09/love-your-monsters/.

Lee, J., & Xu, W. (2018). The More Attacks, the More Retweets: Trump's and Clinton's Agenda Setting on Twitter. *Public Relations Review, 44*(2), 201–213.

Manjoo, F. (2017, May 10). Tech's Frightful Five: They've Got Us. *New York Times*. Retrieved April 5, 2018, from https://www.nytimes.com/2017/05/10/technology/techs-frightful-five-theyve-got-us.html.

Massey, D. (2004). Geographies of Responsibility. *Geografiska Annaler: Series B, Human Geography, 86*(1), 5–18.

Massey, D. (2005). *For Space*. London: Sage.

Massumi, B. (1987). Realer Than Real: The Simulacrum According to Deleuze and Guattari. *Copyright 1*, 90–97. Retrieved August 10, 2018, from http://www.brianmassumi.com/textes/REALER%20THAN%20REAL.pdf.

McLean, J. (2016). The Contingency of Change in the Anthropocene: More-Than-Real Renegotiation of Power Relations in Climate Change Institutional Transformation in Australia. *Environment and Planning D: Society and Space, 34*(3), 508–527. https://doi.org/10.1177/0263775815618963.

McLean, J., & Maalsen, S. (2019). Disrupting Sexism and Sexualities Online? Gender, Activism and Digital Spaces. In C. Nash & A. Gorman-Murray (Eds.), *The Geographies of Digital Sexuality* (pp. 183–202). Singapore: Palgrave Macmillan.

Milan, S., & Treré, E. (2019). Big Data from the South(s): Beyond Data Universalism. *Television & New Media, 20*(4), 319–335. https://doi.org/10.1177/1527476419837739.

Nikitina, S. (2012). Hacker as Trickster of the Digital Age: Creativity in Hacker Culture. *The Journal of Popular Culture, 45*(1), 133–152.

Papacharissi, Z. (2002). The Virtual Sphere: The Internet as a Public Sphere. *New Media & Society, 4*(1), 9–27. https://doi.org/10.1177/14614440222226244.

Parry, D., Johnson, C., & Fullagar, S. (Eds.). (2019). *Digital Dilemmas: Transforming Gender Identities and Power Relations in Everyday Life*. London: Palgrave Macmillan.

Pedersen, S., & Lupton, D. (2018). What Are You Feeling Right Now? Communities of Maternal Feeling on Mumsnet. *Emotion, Space and Society, 26*, 57–63. https://doi.org/10.1016/j.emospa.2016.05.001.

Pesce, M. (2017). The Last Days of Reality. *Meanjin*. Retrieved April 10, 2018, from https://meanjin.com.au/essays/the-last-days-of-reality/.

Pickerill, J. (2003). *Cyberprotest: Environmental Activism Online*. Manchester: Manchester University Press.

Pink, S., Ardevol, E., & Lanzeni, D. (2016). *Digital Materialities: Design and Anthropology*. London: Bloomsbury.

Pile, S. (2010). Emotions and Affect in Recent Human Geography. *Transactions of the Institute of British Geographers, 35*(1), 5–20.

Richardson, L. (2016). Feminist Geographies of Digital Work. *Progress in Human Geography, 42*(2), 244–263. https://doi.org/10.1177/0309132516677177.

Rose, G. (2016). Cultural Geography Going Viral. *Social and Cultural Geography, 17*(6), 763–767. https://doi.org/10.1080/14649365.2015.1124913.

Sedgwick, E. K. (2003). *Touching Feeling: Affect, Pedagogy, Performativity*. Durham, NC: Duke University Press.

Shouse, E. (2005). Feeling, Emotion, Affect. *M/c Journal, 8*(6), 26. Retrieved from http://journal.media-culture.org.au/0512/03-shouse.php/.

Steffen, W., Crutzen, P. J., & McNeill, J. R. (2007). The Anthropocene: Are Humans Now Overwhelming the Great Forces of Nature. *AMBIO: A Journal of the Human Environment, 36*(8), 614–622.

Stephens-Davidowitz, S. (2017). *Everybody Lies: Big Data, New Data, and What the Internet Can Tell Us About Who We Really Are*. New York: HarperCollins.

Whatmore, S. (2002). *Hybrid Geographies: Natures Cultures Spaces*. London: Sage.

# 2

# Framing the More-Than-Real in the Anthropocene

Efforts to confront and engage with environmental dilemmas in different ways abound within the more-than-real. We could read some of these efforts as producing 'weird solidarities' (Gregory 2015) where new forms of sociality are formed from the 'more-than-us-but-not-us' digital spaces. Drawing on Haraway's thinking on solidarity, Gregory (2015) develops the notion of building new relational possibilities through weird solidarities to facilitate staying with the trouble in the Chthulucene/Anthropocene. Weird solidarities are formed across unlikely alliances and with unexpected things, people, communities and institutions. Positionality is crucial in these weird solidarities; Haraway (1998) states that to be anyone we must be one of many, and similarly Gregory states that the sharing economy, and digital platforms more broadly, rely on weird affectual relations to continue. For example, digital platforms have enabled people to be comfortable getting into a car with a complete stranger, thanks to a driver's high rating on a ride-sharing app and its security provisions.

Weird solidarities allow for the presence of data with its material vibrancy (Bennett 2009; Gregory 2015) and so carries with it contentious issues beyond the digital spaces from which they originate.


J. McLean, *Changing Digital Geographies*,
https://doi.org/10.1007/978-3-030-28307-0_2

Connected to these ways of thinking about weird relations, Bennett (2009) frames the material agency of non-human, or not-quite human things, as an important, but often overlooked, part of different worlds. The reasons for doing so are selfish, Bennett writes, as she wants to promote greener forms of human cultures and increase our attentiveness to, and with, non-human and nonliving entities. Such self-centred generosity so honestly stated is compelling and foregrounds an ethics that does not hide behind false altruism. This chapter opens with an analysis of a recent weird solidarity that allegorises debates on the Anthropocene and more-than-real dynamics. It is worth noting at this stage that Shulman (2012) wrote a book called 'More than Real' that offers a history of the imagination in India, examining nondual ways of thinking. The slipperiness of what exists as 'real' and 'unreal' in south Indian philosophies is the focus of Shulman's scholarship—sharing similar conceptual ground with the more-than-real of this book.

## Weird Solidarities in the Digital Anthropocene

Weird solidarities are emerging between academic and activist communities where emotional and affectual responses to changing worlds ground common agendas. As an illustration, we could consider how climate scientists in Australia have produced a video called 'I'm a Climate Scientist' (Hungry Beast 2011) and shared it publicly on YouTube to capture their dissatisfaction with the communication of scientific information about climate change. It begins with a statement:

> In the media landscape there are climate change deniers and believers, but rarely those speaking about climate change are actual climate scientists.

The distinction between real and fake identities is made in the introduction and then the visuals and lyrics of the video go on to identify the multiple ways in which climate science is ignored and why it should be heeded. The video has been watched over 250,000 times. It starts by quoting Gerard Henderson, an Australian conservative journalist, saying 'I'm not a climate scientist' and then launching into a hip hop track with 'Yo, we're climate scientists and there's no denying this…climate change is real!' The names of climate scientists are given in the song, along with their specific areas of expertise and it lists physical processes such as the defrosting of permafrost, increased carbon emissions and feedback mechanisms. A mock-shock jock radio commentator offers that 'The Greenhouse Effect is just a theory sucker' and the scientists call back that 'Yeah and so's gravity/float-away motherfucker!' before ending with call and responses of 'Who's a climate scientist? A palaeontologist? No, a climate scientist!'

The climate scientists' performativity is satirical, humorous and yet also deeply serious. The performers of the song are wearing lab coats and are filmed mimicking the bodily performances of other hip hop artists. The scientists singing in the clip are researchers who work in laboratories and institutions that are identifying a broad range of climate change impact across all the oceans and continents of the world (Intergovernmental Panel on Climate Change 2014), for non-humans and humans. For scientists to participate in an activist-driven hip hop video, not the usual expressive domain for highly trained experts on climate, we can infer that the levels of concern must be quite high and that different ways of communicating the key message were considered necessary.

The 'I'm a Climate Scientist' video attracted an engaged audience: as of August 2018, 2300 people gave the video the thumbs up, compared to 411 offering a thumbs down. I read the efforts of these scientists as a welcome and also weird solidarity. Social media provides an outlet for frustrated scientists to mock and lampoon their opponents and disrupt discourses of 'belief' around climate change and the way that media communicates debates in their field. Drawing on Bennett's vibrant matter is instrumental here as we can see that concerns about the agency of non-human and not-quite human entities are coming together in this performance with epistemological clashes. Despite journalists and mainstream media commentators having no grounding in climate science, these individuals feel comfortable with expressing *beliefs* about the climatic world around them, and experts who do carry such knowledge are motivated to counter these claims in a form of weird solidarity. In Latour's (2014) terms, we can also see a bringing together of unexpected coalitions or blending of elements and institutions that were, at one point in time, more frequently kept separate. The blurring of identities, roles, passions, hopes and ways of engaging with the world, are constitutive of this form of weird and productive solidarity.

The agency of assemblages (Bennett 2009) is evident in the 'I'm a climate scientist' video as affective bodies produce a powerful presence in social media. Rather than offering a calmly lectured retort to political inertia or commentators' climate scepticism, the energy of the video delivers an affective charge. You can't help but feel something in response to the sometimes-awkward gyrations in the clip—mirth and disbelief perhaps, surprise and pleasure too. Playing with tropes of miscommunication of science in mainstream media delivers an intriguing offering of a different way of reading realities.

## Digital Geographies and Digital Anthropocenes

Digital geographies are co-produced by relations between people, technologies and the more-than-human, and include material and immaterial qualities. In taking a non-technological deterministic perspective, I argue that humans engage with the more-than-human world in ways that subvert, challenge and remake humans *and* the more-than-real. Rose (2015, 766) traces the development of geographic work on digital technologies and the tendency for scholars to draw out the agency of *digital* actants, concluding that there is little interest 'in the reflexive and creative agency of human actants'. This research is very interested in the reflexive and creative agency of humans—and has been following Wajcman's (2010) work on feminist technologies as an inspiration for this approach. But it is not just human agency that is of sole interest here: rather, I am also examining the interrelations between human and digital agency that produce such framings as the Anthropocene, or that co-produce digital action that aims to produce changes. By looking broadly at how digital spaces are made, and are remaking us, I want to also decentre humans in discourses and materialities of the digital.

The digital turn in geography has a range of foci and includes work that sits along a spectrum of interest from digital agency to human reflexivity. Ash et al. (2016) describe the digital turn as including geographies through the digital, geographies produced by the digital and geographies of the digital. Geographies through the digital encompass the recognition that digital technologies capture qualitative research and that body of critique on the positivist tendencies of Geographical Imagery Systems. Geographies produced by the digital include exacerbated inequality through social-technological processes that result in a digital divide. Last, geographies of the digital analyse the way human and more-than-human relations are made in digital spaces. As Ash et al. (2016, 8) write, geographies of the digital tend to place it 'as a particular geographical domain with its own logics and structures'. In contrast, I am drawing on Gerbaudo's (2012) approach in talking to transformations stemming from the digital. There are particular qualities of the digital that are important but these are not completely bounded or distinct to other social, cultural and environmental dynamics.

The digital turn is also an opportunity to reconsider conceptualisations of affect and emotion, and how these interact with ways of thinking about, and doing, practices such as activism. Gerbaudo (2012) offers an excellent analysis of the interactions between social media use and offline protest action that considers the dimensions of social relations and emotions in these different forms of public space generated by online action. He argues that the 'emotional quality of the communications of contemporary movements need to be understood in conjunction with the "popular" character of these movements' (Gerbaudo 2012, 161). Emotions play a primary role in our digital geographies and invite further examination. Power circulates through tenuous networks in online movements and the apparent flat structure of online action is belied in Gerbaudo's (2012) research. The way emotion and affect facilitates geographies of digital change forms a key thread in this book.

## Scoping Digital Geographies and Anthropocene (Digital) Power

Far from an exhaustive analysis of the scope of digital geographies, this chapter touches on relevant aspects of this work for the purposes of reimagining the Anthropocene. The massiveness of 'big data' and critiques of the inviolability of the science that aggregates this information is a recurrent theme in geography's digital turn. Elwood (2010) advocates a critical perspective on 'massiveness', arguing that 'there is as yet little theorization of the sociopolitical significance of this growing emphasis on the massiveness of information sets' (Elwood 2010, 353). Massiveness can be seductive in considering sociological meanings from big data but the qualities of digital geographies beg deeper analysis than superficial correlations or associations can suggest. For example, very popular books such as 'Everybody Lies: Big data, little data, and what the Internet tells us about who we really are' (Stephens-Davidowitz 2017) rely on assumptions about intent and meaning from Google searches that are rarely backed up with asking people why they searched what they did. Inferences are at the core of this sort of digital research that builds pictures of 'reality' that hinge upon tenuous assumptions.

A relational view underpins this analysis of changing digital geographies as it is through the bringing together of previously unrelated elements that generates digital action. Müller and Schurr (2016) describe what a relational perspective offers when analysing the similarities—and disjunctures—between Delueze and Latour. They state that a relational view takes the position that 'action results from linking together initially disparate elements' (Müller and Schurr 2016, 217). This builds on Gerbaudo's argument that emotion and affect are crucial in understanding interactions between online and offline action, and my interest in the more-than-real. It is the linking of 'disparate elements' that produces powerful changes, that may be productive or destructive, and as a result the whole becomes more than the sum of its parts. Disparate elements may be individuals, collectives, corporations, environmental forces and immaterial affective flows in digital geographies—producing a dizzying set of contexts and circumstances in social and cultural change processes. The more-than-real accretes a range of elements, bringing together these different actors in relation with each other, and sometimes amplifying the possibilities of each and all.

While the spaces offered by the digital can be damaging, alternative digital sites of resistance and safety also exist. For instance, Shah (2017) describes safe spaces that emerge online when offline realities are not safe and uses the theory of technoaffective frameworks to explain them. In some situations, online places provide networks and support for those who might not be physically safe offline to do the same. The #KissofLove campaign, based in India, is an example of this where people were able to share examples of physical affection in digital spaces that is mostly prohibited in other spaces. Shah (2017, 193) says that the digital is 'the space where the affective, the visceral, the dream-like and the unexpected can happen and the digital becomes a transformative medium that makes it possible'. The unexpected is what we see when mobilisations generate substantive interventions despite the ordinariness of 'liking' a post or joining a hashtag meme. The linking of disparate elements, the surge in recognition of shared anger, shock, bemusement and disappointment, sometimes elevates digital spaces from the mundane to the more spectacular. The more-than-real can be unpredictable and this can bring about surprising opportunities.

Digital geographies bring together multiple scales, from the micro to the macro, the local to the global. We can read the current fascination with the Anthropocene in global environmental thinking through the more-than-real. In Demos (2017) work 'Against the Anthropocene', we are introduced to a short film called 'Welcome to the Anthropocene', a video that summarises and introduces key moments in Anthropocenic thinking. Demos' strong critique of that video, and the Anthropocene thinking it captures, concludes that the Anthropocene rhetoric—joining images and texts—frequently acts as a mechanism of universalisation, which enables the military-state corporate apparatus to disavow responsibility for the differentiated impacts of climate change, effectively obscuring the accountability behind the mounting eco-catastrophe and making us all complicit in it.

According to Demos, the Anthropocene is a spectacle where data has generated an understanding of global environmental change as human-induced (Steffen et al. 2007, 2011). The Anthropocene has multiple framings: as crisis (Hamilton et al. 2015); as rupture (Gibson-Graham and Roelvink 2009); as moment of grief and hope (Head 2016); as scientific cause of concern (Crutzen 2006). Digital tools have enabled us to scale up our understanding of environmental damage and determine that humans are now a geologic force of change. The inevitability of that change is, however, challenged, offering glimpses of points of light in the darkness (Head 2014; McLean 2016). Donna Haraway has problems with the Anthropocene idea but doesn't throw it away: rather, she offers complementary terms to bolster the paucity of language and concepts coming with this notion. As a feminist cyborg theorist and philosopher of human–animal connections, Haraway advocates the Chthulucene, drawing on science fiction figurations. This alternative nomenclature comes with the slogan 'Make Kin Not Babies!' suggesting that we need to engage with all those around us—humans, non-humans and the inanimate—and connect with these rather than produce more to love. Haraway states that 'Right now, the earth is full of refugees, human and not, without refuge' (Haraway 2015, 160). In other words, we've rendered strange, at a global scale, that which nurtures us, and non-humans too.

This is nothing new for Indigenous people, who have experienced the devastation of colonisation and seen their connections with Country challenged again and again. As Whyte (2018, 226) states 'Some Indigenous peoples, then, offer the idea that we confront climate change having already passed through environmental and climate crises arising from the impacts of colonialism'. He offers a compelling argument for centring Indigenous Climate Change Studies as a way to rework and, ultimately, decolonise the Anthropocene (Whyte 2017). Indigenous scholars have countered static framings of culture and the environment from a number of perspectives: anthropological, geographic, sociological and literary. Whyte, a Potawatomi man, challenges the portrayals of post-apocalyptic landscapes as erasing the experiences of Indigenous peoples who are already living with cultural, environmental, social and political devastations as a result of colonialism. The colonial violence that still disturbs Indigenous livelihoods are overwritten by these representations of Anthropocene disaster. Geographer Lesley Head, in her compelling *Grief and Hope in the Anthropocene,* says we need to 're-imagine humans as a force for environmental good, or at least not essentially bad or damaging, as some of the Anthropocene framing suggests' (2016, 11). If we reposition ourselves as having agency, as proponents for, and actants of, doing *more good* than harm, then the possibilities of apocalyptic environmental change might be mitigated. We have to imagine ourselves as capable of doing better before we actually can act in such a way. Similarly, Gibson-Graham and Roelvink (2009) pursue a constructive perspective when advocating a different economic praxis in the Anthropocene. They see this environmental moment as an opportunity to reimagine connections between planetary and everyday life: multiple scales could collide and be reformed (Gibson-Graham and Roelvink 2009).

The ordinariness of environmental dilemmas is also a part of the Anthropocene debate. When arguing that 'crisis is where we live', Houston (2013) calls for more narratives of environmental justice, alongside a focus on the kinds of work that the Anthropocene idea does. Who is interpellated as the 'us' that is humans driving this epoch? Who experiences the costs of neglected environmental degradation in the form of greater vulnerability to intensified storms and droughts? The effects of the Anthropocene are not evenly distributed and not everyone, everywhere is equally responsible for its coming into being. The Global North has benefitted from rapid industrialisation processes, fuelled by colonial empires reaching around the world, while the South shall bear the brunt of greater environmental insecurities. It is here that environmental justice questions emerge.

## Gendering the Anthropocene

Gendered imaginings of the Anthropocene are strong and sometimes elided over, despite the etymology of the term. Anthropocene has been translated as the 'age of man' (Kolbert 2011). A feminist critique from Gibson-Graham (2011, 1) includes the explanation, and almost apology, that 'We have come to see that the scale of the environmental crisis we are part of is creating a new "we" and convening new publics on this planet'. An intersection with the new publics emerging from and within digital spaces is evident here: the new 'we' is not an identity Gibson-Graham (2011) relish. It is with reluctance that they turn from regional development foci to consider this global environmental crisis that is here, now. Indeed, Gibson-Graham (2011) ask us to 'start where you are' and recognise that we are participants in a 'becoming world', drawing on vital materialists such as Bennett (2009) to ground their argument.

Another connection with the discourses of the Anthropocene and digital geographies is found in Gibson-Graham's (2011) argument on how regional development can be supported and facilitated by the vitality of the digital. They hopefully ask 'Might we belong differently now that the vibrant materiality of the internet and open source software allow for new interconnections in a potentially democratized world?' (Gibson-Graham 2011, 10). That materiality of the digital is something geographers are well equipped to engage with, given the integrated approach that geography takes to social, political, cultural and economic issues, where material and discursive matters are usually considered at the same time rather than in isolation.

The Anthropocene idea can be further challenged with respect to its philosophical roots. Returning to Demos' (2017, 35) critique of the Anthropocene, it is possible to reframe as not the age of man, per se, but the 'age of "corporate activities"'. Demos prefers the Capitalocene conceptualisation of the environmental crisis that we're in, arguing that not everyone everywhere is equally responsible for this spacetime. In advocating a counter-Anthropocene moment, Demos challenges those who are dislocating the causes of global environmental change from advanced capitalism to think again.

A strong experimental approach to the Anthropocene is captured in 'Anthropocene Feminism' (Grusin 2017) that offers critiques for masculinist and capitalocentric approaches to global environmental changes (Gibson-Graham 1996). Grusin (2017) edits a vital collection of pieces that address aspects of intersection between feminist and Anthropocenic thinking. In his introduction, Grusin traces the long ecofeminist and feminist lines of thinking that led to this discussion, including Haraway's Cyborg Manifesto. Grusin (2017, x) summarises the approach taken by those contributing to Anthropocene Feminism in the following way:

> In contradistinction to the too often unquestioned masculinist and technonormative approach to the Anthropocene taken by technoscientists, artists, humanists, or social scientists, we created the concept of anthropocene feminism to highlight the ways in which feminism and queer theory might offer alternatives to these approaches.

Feminism has, Grusin and collaborators argue, already anticipated the question of the Anthropocene and in examining what feminism has to say to this idea, and what the Anthropocene does to feminism, we must draw on earlier feminisms as well as looking forward. 'Anthropocene feminism' is put forward as a provocation: an offering to challenge and invite responses. Can we remake the Anthropocene with our feminist knowledges, and how should feminism continue if we take seriously this shifting of human–nature relations?

The spatiality and temporality of the Anthropocene concept has received feminist critique. Scholars such as Di Chiro (2016) argue that it is no accident that narratives of the Anthropocene have come to prominence at the same time as peak neoliberal capitalism extends its reach around the world. The same systems that propel neoliberal agendas underpin a mindset that frames environmental engagement in anthropocentric terms. Dissecting what the Anthropocene means for humans and more-than-humans has distinctly geographic, digital aspects to it. Drawing on Elwood's (2010) analysis of the geoweb and how geographers are exploring societal implications of the digital, the constitution of social relations, bodies and institutional entangled with the Anthropocene shall be analysed in this book. The power relations and governance regimes reproduced in the Anthropocene concept are encapsulated and co-produced by changing digital geographies. The agents of change that I focus on herein are transformative actors in their multiple forms. The Anthropocene could also be conceptualised as more-than-real: contingent, digital and generalised.

There might be hope in the awareness that humanity is globally impacting on the earth and that this will translate into human action to mitigate those effects. Awareness is worth little if effective action does not flow from changed minds. At the same time, the anthropocentrism that is at the heart of the Anthropocene idea needs consideration.

Our self-elevation to agents of geological change has a touch of the narcissistic to it, and can be linked to our modernist tendency to continually place humanity outside of, or above, natural processes (Head 2016). Keeping this in mind, this book considers how digital action is linked to everyday material realities. The binaries that delimit conceptualisations of digital spaces are countered by thinking through the more-than-real. It's useful to see the way we behave online as more-than-real because the circumvention of space and place that the digital allows can warp geographies of responsibility (Massey 2004).

## Processing the Anthropocene in Relation to the More-Than-Real

The contradictory possibilities of the 'more-than-real' (McLean et al. 2016) are present at a global scale with the expansion of access to these spaces, and the deepening reach of the digital in everyday life. Drawing on more-than-human thinking, the 'more-than-real' is an idea that explains the paradoxical ways that digital spaces amplify and collapse geographies, reworking spatial connections and disconnections. Similarly, the Anthropocene is defined and envisioned as a global scale epoch and which is produced by digital ways of knowing; aggregations and 'smoothing-over' of data are required to construct this global scale of environmental change. The processes that come together to form the more-than-real are bound in relation to the same processes of the Anthropocene. Scaling-up of technologies and data from the micro to the macro, glossing over differences at local scales, universalising patterns of access and imagining technologies as end-point solutions are some of these processes.

As Chandler (2017) writes in his analysis of securing the Anthropocene through hacktivism, the new epoch is troubling in its application as short-term, project-based approaches tend to dominate. The gloss and veneer of datafication that ignores histories and geographies of disadvantage and inequalities, such as described in Chandler's analysis of Jakarta as a city of the Anthropocene, makes the epoch itself seem shallow and problematic:

> The lack of temporality of the emergent assemblages of the Anthropocene mean the "what-is-ness" of the world is enhanced by seeing it only as a momentary event, liable to momentary interventions, rather than in terms of long-term problems that need long-term solutions. (Chandler 2017, 125)

The snapshot imaging of the Anthropocene is enabled by the more-than-real, where responsibility and justice are questions that can be obscured through digital tools. Chandler (2017) argues that there is another, more productive, way forward—rather than doing nothing or adopting this short-termism through datafication, policy-makers need to draw on the already existing adaptive capacities of people living in communities such as Jakarta's informal dwellings, or slums, where resilience involves drawing on diverse life-hacks.

In work on islands, resilience and the Anthropocene, Chandler and Pugh (2018) argue that the Anthropocene is now conceptualised with a concretisation of nature/culture binaries. Climate change, a key part of the Anthropocene, is viewed as a problem that requires mitigation, adaptation, controlling or 'fixing', and therefore a 'problem to be faced in the future *rather than as our present condition*' (Chandler and Pugh 2018, 3). The problem with post-human thinkers, Chandler and Pugh (2018) argue, is that they do not extend relationality sufficiently. Calls to embrace animism and recognise Indigenous ways of knowing are a useful start, but they are not an end. We are already in the Anthropocene and yet environmental dilemmas pre-exist this Anthropocene nomenclature. Relationality offers much in the Anthropocene, Chandler and Pugh (2018, 4) argue, but, 'relationality cannot be governmentalised to enable greater control over life'. Relational thinking and action may make the world stranger, rather than more familiar, Chandler and Pugh argue, drawing on Elizabeth Povinelli's (2016) research with Indigenous Australians. There is a limit, they argue, to how we can draw on constructive relational approaches to change the world around us.

It is fitting to spend some time now with Povinelli's geontologies argument and how she is responding to the Anthropocene epoch framing. Povinelli (2016) states that the Anthropocene is a new and controversial proposition since it signals a move from humans affecting the environment to now humans driving geologic change. The shift differentiates humans from each other in a new way where the 'Human emerges as an abstraction on the one side with the Nonhuman world on the other' (Povinelli 2016, 11). Povinelli is arguing that biopolitics, as Foucault conceptualised, needs to be supplemented by an understanding of geontopower. Geologists are having their moment in the cultural, political, environmental and sociological sun. Povinelli states that there is a critical turn towards ontological thinking in the humanities—where assemblage, geontic and 'post-life' critiques are the focus. Geontics or, as Povinelli prefers, geontopower, is 'a mode of late liberal governance' (Povinelli 2016, 17). Expanding on this, Capitalism is opportunistic and animistic in that it sees all entities as potentially able to generate profit. Markets can sell anything in this late liberal age, including air in the form of carbon emissions, water can be disconnected from land ownership and people's organs can attract a price.

Scientific and resource management knowledges and practices have long been critiqued for obscuring and facilitating colonial, racial and gendered power. Shiva (1992, 207) wrote that:

> In every case, the exploitation of nature in the colonies appears to have taken place in two phases. In the first phase, when nature's wealth was considered abundant and freely available, 'resources' were exploited rapaciously. They were not husbanded. In the second phase, once exploitation had created degradation and scarcity, the 'management' of 'natural resources' became important in order to maintain continued supplies of raw material for commerce and industry. So, first land was turned into a resource, then forests and water, and now with the onward march of technology, it is the turn of seeds to be converted into what are called today 'genetic resources'.

Conservation and scientific practices are associated with histories of exploitation and marginalization of Indigenous knowledges. Along these lines, Yusoff (2018) offers a cutting critique of geology as a universalising knowledge that obliterates other ways of knowing and being in 'A Billion Black Anthropocenes or None'. The Anthropocene relies on geologic thinking—an objectivist scientific practice—that can elide histories of racism and obliterate differential patterns of responsibility.

Geontopower, like the Anthropocene concept, carries aspects of the more-than-real too. The contradictions inherent to the creation of humans as abstracted, non-human beings (Povinelli 2016)—despite our biological realities and persistent critiques of nature/culture binaries—where there is no outside to stand in or on, has more-than-real qualities. Familiar geographic and geologic lines and assemblages are blurred and inverted in this Anthropocene moment, and the strangeness that Chandler and Pugh (2018) call to mind abounds. Climate and environmental activists more generally talk about abnormal seasons being the new normal, while scientists are identifying global weirding as a phenomenon (Sweeney 2014). The Anthropocene has more than its fair share of dystopic qualities associated with it and the urgency with which debates around the future of human and non-human life on earth reflects this. Could capitalism also carry more-than-real qualities? In its grand narratives of trickledown economics and the invisible hand of the market, there is more than a touch of the more-than-real. Similarly to the Anthropocene, Capitalism asserts that wealth generation is possible for anyone and at anytime, that histories of extraction and exploitation are not relevant in the here and now; the Anthropocene and Capitalism could be seen as sharing an ahistorical bent that benefits from shallow shiny short-termism. The capitalist system does not necessarily need an individual—you or I—but it is made by us. Such contradictions are the substance of the late liberal age, according to Povinelli, and captured in the geontopower of the Anthropocene. We, humans, are narcissistically naming ourselves as forces of global change through data and evidence that we have gathered and some even believe, including eco-modernists, that we are able to technologically work our way out of such a condition (Asafu-Adjaye et al. 2015).

## Real and Not-Real Digital Geographies

My work on the shape of reality in digital geographies joins the efforts of cultural geographers and critical thinkers who have been at the forefront of feminist and emotional geographic work. Affect and emotion can both propel people to engage with, and produce, social movements, especially as private and public spaces intertwine to produce issue publics (Kim 2012). I want to show how digital geographies, when understood relationally, can include, and has indeed already included, consideration of emotions and affect in a range of contexts.

In the first piece that Soph Maalsen and I wrote on digital geographies (McLean and Maalsen 2013, 252), we grappled with the distinctions between digital and non-digital spaces and didn't escape the not-realing I'm trying to highlight here:

> These two facets, the sort of political action facilitated through social media and personalised politics, seem codependent—it being difficult for one to exist effectively without the other. Together they represent the paradoxical space described by Rose in the encouragement of heterogeneity and the blurring of boundaries between public and private, "real" world and cyber world, challenging the unilocation of person and agency…

We were trying to talk to the remaking of geographic relations and feminist action, remade partly by strategic use of digital technologies, with a case study of substantive digital feminism in Australia. In retrospect, writing about the real should have been a more careful undertaking.

Affect circulates on social media in novel ways: Longhurst (2013) shows that online communication on platforms such as Skype enable greater care possibilities when distance separates mothers from children. Mothers can better assess how their children are when talking with them through videos and stronger relations were reported. Longhurst uses scare quotes when describing the visual aspect of online video mothering as 'seeing', and later when describing the proximity afforded by the digital: 'It is about hearing, and seeing the lived flesh of the "real" person or people on the screen' (Longhurst 2013, 665). Such qualifications could be linked to the framing, most likely unintended, of digital geographies as being other than real.

My work shares the investments and scholarly commitments of geographers who are working on digital sexualities where sometimes the not-realing of digital geographies appears when describing interpersonal, emotional relations. For example, in the compelling introduction to Nash and Gorman-Murray's (2019, 2) notable collection on geographies of digital sexuality, in which I co-authored a chapter with Soph Maalsen, along with statements about how intertwined the digital is with non-digital spaces, there is also a distinction made between digital and real life:

> Intimacy, largely understood as a physical closeness, is being reshaped as people develop genuine connections with others who they may never meet in real life.

It may be necessary to rework what real life is imagined as comprising, then, if genuine intimacy is afforded by digital geographies, as well as what we understand as intimacy being, in and of itself. This book seeks to join these conversations about how the digital works in complex, mediated and contested geographic contexts.

There are instances of not-realing in academic discourse on the digital and its networking possibilities in urban contexts. Castells (2000, 121), for instance, reproduces the real/not-real online/offline binary when discussing sustainability and digital urban initiatives:

> I spent the day with the people who are doing digital cities and it's really a wonderful programme in terms of the relationship to the citizens, how they connect, how they organize themselves, how it interacts, the digital city and the real city, in a very meaningful way with all kinds of political pay-offs.

Castells is bringing the digital into conversations about sustainability in this speech, available as a transcript in *City*, but also simultaneously distinguishing between these two spaces and rendering the digital as not-real.

The distinction between real and not-real spaces is frequently made at the same time as these spaces are pointed out as intertwined. Farries and Sturm (2018) argue that women's rights are infringed upon by digital misogyny and that 'call-out' culture is insufficient to rectify these. In the abstract for their article, they state that 'Supporting creativity also helps to protect against the reproduction of gendered harms, from the real world to virtual spaces' (Farries and Sturm 2018, 1). The problems that surround the prosecution of sexist behaviour in digital spaces are captured in the delineation of the real world from virtual spaces: we frequently think of digital offences as less tangible and less real than others. If we reposition digital spaces as a part of the real world, the political power of activities generated therein may be framed differently as well.

Useful efforts to talk to the dualities created when thinking of digital and non-digital spaces sometimes inadvertently 'not-realed' the former. For instance, work by Madge and O'Connor (2005) starts with an important note on the intersections between digital and non-digital spaces:

> In this paper we use the term cyber/space rather than cyberspace and geographical space, to try to capture the intersecting and simultaneous nature of the virtually real and the actually real.

The 'virtually real' and the 'actually real' are not challenged here—Madge and O'Connor are critiquing the duality between the two domains rather than the conceptualisation of what is real or not. Similarly, popular commentary on the changing digital spaces that we engage with daily continues to (re)present their not-realing.

If we consider the digital as more-than-real, it may help to illuminate the circumvention of space and place that occurs therein, which can warp geographies of responsibility (Massey 2004). Massey (2004) analysed how politics shapes relations of responsibility and examined the question of the stranger without, querying geographic work that pits place in opposition to space: Massey asks us to reconsider the stereotyping of place as real and grounded while space is framed as elusive and intangible. Massey (2004, 7) wrote:

> A regular litany of words accompanies the characteristic evocation of place; words such as "real", "grounded", "everyday", "lived". They are mobilised to generate an atmosphere of earthiness, authenticity, meaning.

Like the above framing of place/space, there is a tendency in everyday conversations and commentary about the digital to frame digital geographies as not real, or immaterial, rather than potentially substantive, having tangible qualities and relational. I draw on Massey's critique of an inward-oriented sense of place in exploring the binary thinking that relates to the not-realing of the digital (Duggan 2017). Kinsley (2014), Duggan (2017) and Ash et al. (2016) have provided key contributions to dismantling digital dualisms and this chapter builds on those by considering the real/not-real binary that sometimes emerges in academic and non-academic contexts. Of course, the more-than-real political strategy can risk producing another binary—this time, between the real and more-than-real. Just as the more-than-human has broadened how geographers engage with the agency of non-humans, perhaps the more-than-real could broaden digital geographic thinking too. Again, this sharing of examples where the digital is framed as other than real is only to outline some epistemologies pertinent to reality in geographic thinking rather than attribute fault. The purpose of this section is not to allege mistaken use of terms for engaging with digital and other spaces, but to point out the ways in which social sciences and humanities thinking has tended to shape perceptions of digital geographies as other than real. These tendencies are not unique to geography as they stem from broader social and cultural practices relating to digital spaces.

## Conclusion

The shimmering facades of the digital and the Anthropocene are intertwined in increasingly compelling and alarming ways. The Anthropocene epoch is produced by digital realities and inflects digital geographies in a two-way exchange. Meanwhile, weird solidarities are forming to counter problematic arguments against climate action and other species extinction threatening processes. Egregious efforts to discredit science that has established evidence for the changing climate, species collapse and other global environmental system pressures are facilitated by digital spaces. But digital geographies also house the rapping climate scientists that introduced this chapter, a good example of how weird solidarities are making alternative interventions that might move the narratives of business-as-usual politics and corporations. Scientists and political satirists felt compelled to make a rap video to send up the beliefs of those countering climate science facts. They offer the satire in a deeply serious way to find traction in the loose debates on climate change and elicit different affect in the process.

The linking of disparate elements, the surge in recognition of shared anger, shock, bemusement and disappointment, sometimes elevates digital spaces from the mundane to the more spectacular. The more-than-real can be unpredictable and this can bring about different opportunities. Overly limiting categorisations of real/not-real and human/natural are morphing, in part, due to digital Anthropocenes. What we once took for granted as fact-based seems less reliable, while humans and non-humans are finding avenues of intervention in creative channels. I suggest that the more-than-real concept could be helpful here, as a way to differently navigate the shimmering and troubling facades that are part of digital geographies. At the same time, different ways of belonging are facilitated by the digital and some are using these paths to remake the Anthropocene. Simultaneously, the Anthropocene is produced in part by the digital—as a concept, as a geologic claim and as a rupture. To summarise, the more-than-real offers a way to rethink these categories and to understand surprising and concerning changes happening in our worlds.

The fracturing of universal thinking that underpins the Anthropocene has been helping peel back the certainties in this worrying new epoch. Povinelli and Yusoff have given new readings of geological thinking and imagining: geontopower offers a reading of the wrangling already underway in smoothing over the responsibility of environmental crises, while the demand to foreground multiplicity in reworlding the Anthropocene is generative. Thinking of digital geographies in light of these offerings will allow space for bringing together claims of, and work towards, environmental and social justice.

## References

Asafu-Adjaye, J., Blomquist, L., Brand, S., Brook, B. W., DeFries, R., Ellis, E., et al. (2015). *An Ecomodernist Manifesto*. Retrieved from http://www.ecomodernism.org/.

Ash, J., Kitchin, R., & Leszczynski, A. (2016). Digital Turn, Digital Geographies? *Progress in Human Geography, 42*(1), 25–43. https://doi.org/10.1177/0309132516664800.

Bennett, J. (2009). *Vibrant Matter: A Political Ecology of Things*. Durham, NC: Duke University Press.

Castells, M. (2000). Urban Sustainability in the Information Age. *City, 4*(1), 118–122. https://doi.org/10.1080/713656995.

Chandler, D. (2017). Securing the Anthropocene? International Policy Experiments in Digital Hacktivism: A Case Study of Jakarta. *Security Dialogue, 48*(2), 113–130.

Chandler, D., & Pugh, J. (2018). Islands of Relationality and Resilience: The Shifting Stakes of the Anthropocene. *Area*. https://doi.org/10.1111/area.12459

Crutzen, P. J. (2006). The "Anthropocene". In E. Ehlers & T. Krafft (Eds.), *Earth System Science in the Anthropocene*. Berlin: Springer.

Demos, T. J. (2017). *Against the Anthropocene: Visual Culture and Environment Today*. Berlin: Sternberg Press.

Di Chiro, G. (2016). Environmental Justice and the Anthropocene Meme. In T. Gabrielson, C. Hall, J. M. Meyer, & D. Schlosberg (Eds.), *The Oxford Handbook of Environmental Political Theory* (pp. 362–383). Oxford: Oxford University Press.

Duggan, M. (2017). Questioning "Digital Ethnography" in an Era of Ubiquitous Computing. *Geography Compass, 11*(5). https://doi.org/10.1111/gec3.12313.

Elwood, S. (2010). Geographic Information Science: Emerging Research on the Societal Implications of the Geospatial Web. *Progress in Human Geography, 34*(3), 349–357. https://doi.org/10.1177/0309132509340711.

Farries, E., & Sturm, T. (2018). Feminist Legal Geographies of Intimate-Image Sexual Abuse: Using Copyright Logic to Combat the Unauthorized Distribution of Celebrity Intimate Images in Cyberspaces. *Environment and Planning A: Economy and Space*. https://doi.org/10.1177/0308518x18786964.

Gerbaudo, P. (2012). *Tweets and the Streets: Social Media and Contemporary Activism*. London: Pluto Press.

Gibson-Graham, J. K. (1996). *The End of Capitalism (As We Knew It): A Feminist Critique of Political Economy*. Minneapolis: University of Minnesota Press.
Gibson-Graham, J. K. (2011). A Feminist Project of Belonging for the Anthropocene. *Gender, Place and Culture, 18*(01), 1–21.
Gibson-Graham, J. K., & Roelvink, G. (2009). An Economic Ethics for the Anthropocene. *Antipode, 41*(1), 320–346.
Gregory, K. (2015, February 1). *Weird Solidarities*. Retrieved from http://dx.doi.org/10.2139/ssrn.2570492
Grusin, R. (Ed.). (2017). *Anthropocene Feminism*. Minneapolis: University of Minnesota Press.
Hamilton, C., Gemenne, F., & Bonneuil, C. (Eds.). (2015). *The Anthropocene and the Global Environmental Crisis: Rethinking Modernity in a New Epoch*. London: Routledge.
Haraway, D. (1998). Situated Knowledges: The Science Question in Feminism and the Privilege of Partial Perspective. *Feminist Studies, 14*(3), 575–599.
Haraway, D. (2015). Anthropocene, Capitalocene, Plantationocene, Chthulucene: Making Kin. *Environmental Humanities, 6*(1), 159–165. Retrieved November 1, 2017, from https://read.dukeupress.edu/environmental-humanities/article/6/1/159/8110/Anthropocene-Capitalocene-Plantationocene.
Head, L. (2014). Contingencies of the Anthropocene: Lessons from the 'Neolithic'. *The Anthropocene Review, 1*(2), 113–125.
Head, L. (2016). *Hope and Grief in the Anthropocene*. London: Routledge.
Houston, D. (2013). Crisis Is Where We Live: Environmental Justice for the Anthropocene. *Globalizations, 10*(3), 439–450.
Hungry Beast. (2011). *I'm a Climate Scientist*. Video Recording, YouTube. Retrieved July 20, 2018, from https://youtu.be/LiYZxOlCN10.
Intergovernmental Panel on Climate Change. (2014). *Climate Change 2014: Impacts, Adaptation, and Vulnerability: Summary for Policymakers*. Retrieved April 20, 2019, from http://ipcc-wg2.gov/AR5/images/uploads/IPCC_WG2AR5_SPM_Approved.pdf.
Kim, Y. M. (2012). The Shifting Sands of Citizenship: Toward a Model of the Citizenry in Life Politics. *The Annals of the American Academy of Political and Social Science, 644*, 147–158.
Kinsley, S. (2014). The Matter of 'Virtual' Geographies. *Progress in Human Geography, 38*(3), 364–384.
Kolbert, E. (2011). Enter the Anthropcoene—Age of Man. *National Geographic, 219*(3), 60–61, 64–65, 69–73, 75–77, 79, 81, 83, 85.

Latour, B. (2014). Love Your Monsters. *Next Nature.* Accessed December 4, 2018, from http://www.nextnature.net/2014/09/love-your-monsters/.

Longhurst, R. (2013). Using Skype to Mother: Bodies, Emotions, Visuality, and Screens. *Environment and Planning D: Society and Space, 31*(4), 664–679. https://doi.org/10.1068/d20111.

Madge, C., & O'Connor, H. (2005). Mothers in the Making? Exploring Liminality in Cyber/Space. *Transactions of the Institute of British Geographers, 30*(1), 83–97.

Massey, D. (2004). Geographies of Responsibility. *Geografiska Annaler: Series B, HumanGeography, 86*(1), 5–18.

McLean, J. (2016). The Contingency of Change in the Anthropocene: More-Than-Real Renegotiation of Power Relations in Climate Change Institutional Transformation in Australia. *Environment and Planning D: Society and Space, 34*(3), 508–527.

McLean, J., & Maalsen, S. (2013). A Geography of Revitalised Feminism. *Geographical Research, 51*(3), 243–256. https://doi.org/10.1111/1745-5871.12023.

McLean, J., Maalsen, S., & Grech, A. (2016). Learning About Feminism in Digital Spaces: Online Methodologies and Participatory Mapping. *Australian Geographer, 47*(2), 157–177. https://doi.org/10.1080/00049182.2016.1138842.

Müller, M., & Schurr, C. (2016). Assemblage Thinking and Actor-Network Theory: Conjunctions, Disjunctions, Cross-Fertilisations. *Transactions of the Institute of British Geographers, 41*(3), 217–229. https://doi.org/10.1111/tran.12117.

Nash, C., & Gorman-Murray, A. (Eds.). (2019). *The Geographies of Digital Sexuality*. Singapore: Palgrave Macmillan.

Povinelli, E. (2016). *Geontologies: A Requiem to Late Liberalism.* Durham, NC: Duke University Press. https://doi.org/10.1215/9780822373810

Rose, G. (2015). Rethinking the Geographies of Cultural 'Objects' Through Digital Technologies: Interface, Network and Friction. *Progress in Human Geography, 40*(3), 334–351.

Shah, N. (2017). The Cup Runneth Over: The Body, the Public and Its Regulation in Digital Activism. *Crime, Media, Culture, 13*(2), 187–198. https://doi.org/10.1177/1741659017701803.

Shiva, V. (1992). Resources. In W. Sachs (Ed.), *The Development Dictionary: A Guide to Knowledge as Power* (pp. 206–218). Johannesburg: Witwatersrand Press.

Shulman, D. (2012). *More Than Real: A History of the Imagination in South India*. Cambridge: Harvard University Press.

Steffen, W., Crutzen, P. J., & McNeill, J. R. (2007). The Anthropocene: Are Humans Now Overwhelming the Great Forces of Nature. *AMBIO: A Journal of the Human Environment, 36*(8), 614–622.

Steffen, W., Grinevald, J., Crutzen, P., & McNeill, J. (2011). The Anthropocene: Conceptual and Historical Perspectives. *Philosophical Transactions of the Royal Society A: Mathematical, Physical and Engineering Sciences, 369*, 842–867. https://doi.org/10.1098/rsta.2010.0327.

Stephens-Davidowitz, S. (2017). *Everybody Lies: Big Data, New Data, and What the Internet Can Tell Us About Who We Really Are*. New York: HarperCollins.

Sweeney, J. A. (2014). Command-and-Control: Alternative Futures of Geoengineering in an Age of Global Weirding. *Futures, 57*, 1–13.

Wajcman, J. (2010). Feminist Theories of Technology. *Cambridge Journal of Economics, 34*(1), 143–152. https://doi.org/10.1093/cje/ben057.

Whyte, K. (2017). Indigenous Climate Change Studies: Indigenizing Futures. *Decolonizing the Anthropocene. English Language Notes, 55*(1), 153–162.

Whyte, K. P. (2018). Indigenous Science (Fiction) for the Anthropocene: Ancestral Dystopias and Fantasies of Climate Change Crises. *Environment and Planning E: Nature and Space, 1*(1–2), 224–242.

Yusoff, K. (2018). *A Billion Black Anthropocenes or None*. Minneapolis: University of Minnesota Press.

# 3

# Digital Action, Human Rights and Technology

## Dialogues on Human Rights and Technology

The more-than-real is remade in particular places and with specific human–technology relations. To get a sense of one such remaking, this chapter analyses a one-day conference on 'Human Rights and Technology' in Sydney that I attended in July 2018, organised by the Australian Human Rights Commission (AHRC). The AHRC is an independent statutory organisation that is accountable to the federal government through the Attorney General.[1] The conference was held at the Four Seasons Hotel, an expensive hotel on the Sydney Harbour, and facilitated the launch of the AHRC's 'Human Rights and Technology Issues Paper'. I was not sure if I could attend, given the relatively high cost of the event—AUD 275 for 'Community Tickets' and AUD 495 for others. But I was also invited to attend the Roundtable on Human Rights and Technology at the AHRC the day after the Conference

[1]The AHRC was established in 1986 as Australia's national human rights peak organization. The Commission educates the public about, and responds to complaints under, the *Racial Discrimination Act* (1975), the *Sex Discrimination Act* (1984), and the *Disability Discrimination Act* (1992), amongst other roles.



J. McLean, *Changing Digital Geographies*,
https://doi.org/10.1007/978-3-030-28307-0_3

and wanted to find out what the Commission saw as the main issues shaping the debate human–technology relations and rights. I was also unsure about the conversations that might emerge around human rights and technology: would we be discussing *digital* technologies, as broadly defined as that term is, given that the promotional material for the event shared images of robots and binary code? Or all technologies? And what rights would form the focus of the conversation?

Hints of the emphasis for the (digital) technological part of the event emerged in the Conference programme and sponsorship. The Conference and Issues Paper that was launched that day was part of a larger project on Human Rights and Technology that the AHRC had begun earlier in 2018. This AHRC project has a range of partners including the federal government's Department of Foreign Affairs and Trade, Herbert Smith Freehills (a corporate law firm), LexisNexis (a legal database corporation), and the University of Technology Sydney. The conference itself was sponsored by Cisco 'the worldwide leader in IT and networking' (Cisco, n.d.). It was unlike most conferences that I usually attend, in terms of the structure and the evident money behind it, but also in terms of accessibility provisions throughout. Captioning of presenters on a large screen enabled audio accessibility and Auslan interpreters worked throughout the event to translate talks, questions and answers; the accessibility by design (Iwarsson and Ståhl 2003) aspect of the conference was highly appreciated and applauded by the participants. Question time was limited to digitally noted questions that were voted on by attendees and then selected by chairs for the panel or speaker to respond to—there were no 'open' sessions. Food and beverages were in abundance at this event, perhaps unsurprisingly given that it was held in the richest part of the richest city in Australia, which also ranks as the ninth richest city in the world (Pash 2018).

The day began with a Welcome to Country by Uncle Charles 'Chicka' Madden, Gadigal elder, who encouraged everyone to feel at ease on Gadigal lands and waters, with warmth and generosity. It always strikes me how hospitable Indigenous elders are in Welcome to Countries, when injustices continue to characterise settler colonial Australia (Langton and Mazel 2008; Porter 2010). Uncle Charles' welcome was followed by a short video from Julie Bishop, the then federal Minister for Foreign

Affairs, who said that human rights apply online just as they do offline, and that technological advances are disrupting everyday lives. Technologies are providing new opportunities, Bishop said, and we now have the 'whole world open to us' thanks to disruptive technology. The emergence of the disruption theme, sometimes without critique, was evident in this opening talk and it reappeared throughout the day in various guises, reflecting the omnipresence of disruption thinking (Yu and Hang 2010).

Ed Santow, the Human Rights Commissioner at AHRC since 2016, championed the Human Rights and Technology project and gave an agenda-setting introduction to the conference. He spoke about how we need to use a 'responsible innovation' approach when adopting new technologies and that 'technology should serve humanity…whether it does will be dependent on the values we insist on'. The Human Rights Commissioner gave the example of TayTweets (@TayandYou), an Artificial Intelligence (AI) Twitter account that Microsoft launched that responded to input from Twitter users. As part of its algorithm, a 'repeat after me' function played out in a disturbing way as racist, sexist and white supremacist comments were reproduced (Kleeman 2016). The bot was shut down within a day of its launch due to the obscene and inflammatory comments it was generating (Neff and Nagy 2016). Santow described how there were only 16 hours between sweet enthusiastic engagement with the bot and vile, hate-filled rhetoric.

Following Santow's introduction, a short video by the World Economic Forum was played, offering a compilation of unnamed Indigenous perspectives, from around Australia, sharing their perspectives on digital technologies. The conference learnt that the AHRC and the World Economic Forum were working on a white paper on technology and human rights that is due for circulation in 2019. I wonder if Indigenous people were involved in its production; the presentation of nameless, placeless, Country-less Indigenous peoples was a deeply settler colonial moment.

Dr. Alan Finkel, Chief Scientist of Australia, gave the next keynote on 'What kind of society do we want to be?'[2] The focus in Finkel's

[2]The full transcript of this talk is available at https://www.chiefscientist.gov.au/wp-content/uploads/Human-Rights-and-Technology.pdf.

talk was on engaging with, and regulating, AI, rather than a broader spectrum of digital technologies, and this is the first hint of the dominance of the AI theme for the day. Finkel drew on examples of earlier technologies and how society has responded to waves of innovation, including in vitro fertilisation (IVF). This technology has transformed reproductive lives around the world and Australia was a leader in responding to this new possibility. Finkel (2018, 5) stated that:

> People thought that it (IVF) was unnatural. That the babies would be deformed or somehow less than fully human. Or that we would start making humans in batch lots, in factories.

But rather than respond in alarmist ways, Australian scientists and medical practitioners started keeping records on children born following IVF conception and established solid regulatory frameworks in place for this technology. The same, Finkel argued, should be pursued for AI. Finkel argued that we could have a mark for digital products, akin to the Fairtrade certification that indicates food products that have been produced ethically including due recompense to the producers, which would recognise ethical technology vendors.

The argument for stronger regulation formed a key theme throughout the Chief Scientist's keynote. On occasion, the capacity of governments to effect this regulation is somewhat inflated and even assumed. For example:

> Governments decide how companies are allowed to use data. Governments decide how to invest public funds in AI development. Governments decide how they want to harness AI, for policing and healthcare and education and social security – systems that touch us all. (Finkel 2018, 5)

Governments do set rules for processes such as the collection of metadata and its distribution but quite quickly control over such information is limited, willfully or otherwise. Within Australia, the Federal Government has been asking telecommunication companies to collect metadata since 2017 on who people text, where they make and take calls, and who emails are sent between (McLean and Mackenzie 2019). Clarke (2018) reports that telecommunication companies are receiving more than 1000 requests for metadata each day, and that these are coming from more than the original agencies authorised to access that information. Such interventions suggest that government control over digital technologies is not absolute and is, quite frequently, let go in predictable ways, while at other times regulation is a force for poor outcomes rather than beneficial ends.

During the conference, the Federal Government's mishandling of the process of creating digital health records for Australian citizens, the 'My Health Records' initiative, was held up as an instance of the problems that emerge when governments engage poorly with digital technologies. The Twitter discussion that was happening during and around the conference, and was being drawn on for questions to panel and keynote presenters, used the hashtag #techrightsau. A search of the 4427 tweets that use this hashtag shows that 70 tweets also included #MyHealthRecord. For example, Prof Kim Weatherall tweeted the following about the need for stronger human rights protection associated with a digital initiative (Fig. 3.1).

The concern from privacy groups, academics, individuals and some politicians about the compulsory nature of the My Health Record initiative forced the Federal Government to introduce an opt out path for those who do not wish to have such records kept.

**Fig. 3.1** Prof Kim Weatherall's tweet about #MyHealthRecord and human rights from the Human Rights and Technology Conference

Transforming AI to work better, and for more people, was a central message from Kathy Baxter, author and 'Ethical AI Practice Architect' at Salesforce, who was next to offer a keynote at the conference. AI must be used carefully, Baxter argued, as it can amplify bias and is generally deemed plausible—due to 'tech worship' (Fig. 3.2).

Baxter discussed the way that AI reproduces, and concentrates, bias: 'AI is not neutral, it is a mirror that reflects back on us'. When I tweeted this at the conference, Professor Deb Verhoeven, from University of

**Fig. 3.2** Screenshot of the AHRC tweeting Kathy Baxter at the Human Rights and Technology conference

Technology Sydney (@bestqualitycrab), responded that it was more than a mirror, AI is a distortion of a narrow segment of the world rather than an all-encompassing, reflective device. The distortion that AI does is a manifestation of the more-than-real qualities of digital geographies. Rather than simply mirroring back how the world works, AI will *magnify* certain assumptions and prejudices, possibilities and constraints, that the makers of that device or tool allow. AI is involved, actively, in worlding and distorting (Fig. 3.3).

One of the most controversial statements of the conference came from Professor Toby Walsh, an academic working on AI at the University of New South Wales, who said that 'hopefully AI will do the 4 Ds'. The '4 Ds' were defined and critiqued by conference participant Dr Fiona Martin (Fig. 3.4).

The question of the problems around scaling such work and overseeing the reach of AI is crucial. When we see the ramifications of the mistakes of AI in action, these questions that drill down into techno-optimist framings are demonstrably even more important. At the same time, the statement that 'dirty, dull, difficult and dangerous' tasks should be overtaken by technological interventions carries assumptions and judgments about what work means, and how it is valued. To overlook these and somewhat flippantly suggest that they should be transformed reveals values and hierarchies about what work is and needs to involve. Further, the problems that come with unrestrained automation are already evident, such as in the #notmydebt case study that I discuss in Chapter 4, detailing the robodebt recovery of so-called debts owed by welfare recipients with Australian Taxation, plays out some of these concerns. The mistakes that the system produced resulted in 4000 of these 20,000 automated decisions being wrong (Tsoundarou 2017; Galloway 2017).

The conference progressed through panels on 'Human Rights by Design: Responsible Innovation' and Professor Genevieve Bell's keynote 'Food for thought: how to be human in the digital world'. Bell comes from the tech industry, having worked at Intel for over two decades, bringing her anthropological nous to computer technologies. Her insights on the power of algorithms to always reproduce pre-existing bias and histories of exploitation were powerful:

**Fig. 3.3** Twitter exchange between Prof. Deb Verhoeven and me on how AI works

**Fig. 3.4** Dr. Fiona Martin challenging the '4 Ds' at the Human Rights and Technology Conference

> If we have algorithms that can't help but reproduce the past… what does it mean to develop an algorithm that leads to wonder, surprise, or challenge? (Bell 2018)

As a futurist, Bell is keen to push conversations towards this speculative realm, where the possibilities of better digital technologies, that are more human-oriented, become the norm. Bell introduced her talk with an image of 'The challenge of cyber physical systems' (Fig. 3.5). This image captured four 'revolutions' in technological production, moving from the first Industrial Revolution to what could be conceptualised as the Fourth Revolution today with 'Cyber Physical Systems' (Industry 4.0). Bell rhetorically asked the audience what is missing in the image and exclaimed 'People!'. The worlding that is imagined with these revolutions is delineated as devoid of environmental realities, too.

The paradox here is that while, as Bell says, 'Data is always retrospective', digital technologies are frequently viewed as revolutionary and forward-looking, providing new possibilities and conveniences. That assumption is wrong, at least in terms of where the coding comes from to inform these technologies, but Industry 4.0 carries that tension with it, an implicit paradox that could be read as a part of the more-than-real. The paradox can surprise us when technologies fail to revolutionise ways of beings and instead reproduce inequities, biases, assumptions

**Fig. 3.5** Photo of Prof. Genevieve Bell and the revolutions from the Human Rights and Technology Conference (taken by author)

and stereotypes, as the @TayandYou chatbot did. Rather than opening up different ways of connecting and being together, Tay was trolled and recreated the worst aspects of digital spaces in a short period of time and in an extreme form. More-than-real spaces are tricky.

One of the best panels of the day was 'Inclusion riders: People with disabilities and technology' that included Emma Bennison, Executive Officer of the Blind Citizens Australia, Manisha Amin, Chief Executive Officer of the Centre for Inclusive Design and Alastair McEwin, Disability Discrimination Commissioner within the AHRC. Key themes of the discussion included bringing people into the room who are not already included—Manisha Amin described this as creating a 'virtuous circle rather than a vicious cycle by taking ideas from the margins and making them central'. This panel was shared as a podcast on the Australian Broadcasting Corporation's (ABC) website (ABC 2018).

Emma Bennison and Alastair McEwin talked about the strengths of digital technologies that provide visual and audio support for navigating communication and mobility, including 'Be my eyes' and 'Aira'. These technologies use augmented reality to assist people navigating unfamiliar spaces: 'Aira connects people who are blind or low vision to a trained professional agent who is dedicated to further enhancing their everyday experience—completely hands-free assistance at the touch of a button' (Aira 2018). Emma gave the example of a lift that was not accessible and she was able to call a service in the United States that acted as her 'eyes' and let her then use the lift to meet her friend. These tools are game-changing from Emma's perspective:

> We talked before about the ability of technology to create an emotional response and I have to say that probably not since the time that I got my first personal computer in 1988 have I experienced such a strong emotional response to a piece of technology as I have to this, because it really does spell liberation for me. I was able to, the other day, come out of an unfamiliar building, go outside and have a coffee, I was able to call up Aira, and navigate to the coffee shop. It might sound really simple to you but for me it was just incredibly liberating. I have found myself… and others in the community … who are just brought to tears by this technology. (transcribed from ABC 2018)

The independence that these augmented reality tools gives to those who would have had to rely on other supports or people clearly empowers and enables people like Emma. The generative potential of digital spaces is clear here. The capacity to move freely in novel spaces brings strong emotional responses of relief and joy for those who previously had to fight for this. Emma goes on to express concern that these new technologies will make advocates and corporations complacent about introducing accessibility to all tools, devices and facilities.

The problems with governance and bureaucracy when thinking through the most appropriate ways to support individuals and communities with new technologies were discussed in this 'Inclusion rider' panel. The panel raised serious issues associated with the work of the National Disability Insurance Agency (NDIA), the Australian statutory

Federal agency charged with implementing the National Disability Insurance Scheme (NDIS), and the strictures it has in place for funding aids and supports for those with disabilities. The NDIA does not allow people to purchase mainstream technologies, such as iPhones, to use as devices under their funding model, as these are not defined as disability supports. The demarcation of particular technologies as 'for disability use' or otherwise is concerning and not cost-effective. For example, mainstream technologies such as iPhones are less expensive than the specialist devices that people are encouraged to purchase through the NDIS. During an interview with Digital Gap Initiative founder, this claim was corroborated (October 2018).

Later in the conference, Aza Raskin, ex-Mozilla technologist, now co-founder of the American based Center for Humane Technology, spoke about the debt he felt that he owed for creating infinite scrolling and how he feels that he 'needs to atone for it!' Aza argued that there are a couple of companies controlling information around the world and they want something from us all: attention. The website for the Center for Humane Technology includes a description of 'The Problem' with digital technologies and their mission in the following way:

> Technology that tears apart our common reality and truth, constantly shreds our attention, or causes us to feel isolated makes it impossible to solve the world's other pressing problems like climate change, poverty, and polarization. No one wants technology like that. Which means we're all actually on the same team: Team Humanity, to realign technology with humanity's best interests. (Center for Humane Technology 2018)

An issue with the Center's 'Problem' setting is that they posit that there ever was one 'common reality and truth'. Diverse ontologies have always existed in the worlds that we inhabit and multiple ways of knowing and being need to be recognised, such as Indigenous ontologies (for example Bawaka Country et al. 2018) and multispecies ways of being (Van Dooren and Rose 2012). The overall goal of an institution like the Center for Humane Technology to realign technology in humanity's best interest is salutary but as the discussion above shows about disability, inclusion and technologies, there is no singular 'best interest' now, and there never has been. Just as there is no singular reality. Rather, we experience the worlds we exist in and create and challenge in multiple ways at all times, and to start from the position of flattening that complexity sets you up to fail.

Aza asked the audience to participate in a quick activity to highlight the issues of digital technologies and our everyday lives during his talk. He asked us to open the photo app of our smartphones and hand them over to the person sitting next to us. I was about to do this when the man next to me said 'No, that's the whole point of the activity, we're not supposed to want to do this'. I laughed and thought about why I was belying the audience participation premise: as a mother, my phone is nearly always available to my child and there is not anything on the photo app that I would not be okay with him seeing. So the position that Aza was coming from was quite different to the lived experience that I have with digital spaces. I experience privacy dilemmas differently to the man who sat next to me, just as others may. And this again underlines the critique of the 'common reality' mission—we are patently not all using digital technologies in the same ways, in the same spaces, at the same times.

In the conclusion of the Conference, Human Rights Commissioner Ed Santow stated that throughout the day, we have identified the problems and opportunities associated with technology, and argued that we now need to find solutions. Putting forward that there is nothing inherently good or bad about technology, he used the example of a knife to explain seeming contradictions in dual-use technology: knives are used functionally every day to cut food so we can eat it, but they can also be used to stab someone. The President of the AHRC, Rosalind Croucher, gave her overview of the key issues of the day and suggested that we are in the middle of a 'new digital revolution'. Adapting the slogan of the French Revolution, the President said that we can have the following 'keywords' for the new digital revolution: 'Security, responsibility, integrity, inclusivity, humanity. And without humanity we will be abandoning all hope in entering this new revolution!'

On Twitter as the day was coming to a close, I wondered if sustainability could be added to the five keywords that the President of the Commission offered to summarise and frame the event. Throughout the proceedings, the best catering was provided in the luxuriously appointed Conference space; you could smell and taste the money at this gathering. Canisters of water were provided in the foyer, graced with lemon slices. And yet water bottles were also on each table and, halfway through the day, someone came around and replaced even half-used water bottles with brand new ones. The tables and chairs were dressed as if ready for a formal white wedding and the water bottles, pens and notepads at each setting were emblazoned with the Four Seasons logo. Throughout the event, issues relating to sustainability of technology were not mentioned, and the way the Conference itself ran suggested that sustainability was not high on the agenda.

This overview of a day in Australia's human rights and technology discourse and practice sets the scene for how the main themes play out in this book. How do digital technologies intersect with human rights and ethical questions more generally? And where does sustainability, however this is defined, sit in relation to human rights and other rights in digital spaces? The sidelining of most sustainability issues could be attributed to a too-busy, too-crowded discursive space but that does not adequately address the matter. Sustainability discourses have circulated for over 30 years since the Brundtland Report brought it to a wider audience (Sneddon et al. 2006). Some technology firms are addressing sustainability—even if tokenistically—and yet we do not see these concerns breakthrough to broader critiques of technological shifts.

In the Human Rights and Technology Issues Paper that was launched at the conference, sustainability is mentioned in passing:

> Economic inequalities may emerge in the application of technology (for example, displacement of the labour force through robotics), and through market effects of technology (for example, displacement of small competitors, power concentration, price discrimination, value chain control). Economic inequality has consequences for individual and communal participation in social, cultural and political life.
>
> Conversely, new technologies can reduce inequality and enable participation for those who have been traditionally excluded. For example, sustainable energy technologies can dramatically improve the lives of

> citizens in developing countries and ameliorate the impacts of climate change. (AHRC 2018, 16)

Here, technologies are a solution rather than conceived of as part of the dilemmas we face in increasing consumption of rare metals and minerals, the production of carbon emissions, and the disposal of e-waste. Further, the conceptualisation of technology as economically contingent suggests an engagement with ideas around sustainability that requires further examination. I wonder what different sorts of questions and dialogues might have emerged if we explicitly brought these conversations together and whether these might bring about varied questions and conclusions. The gaps in the dialogue may have resonated, opening up more questions about how we should and could be living differently with digital technologies.

## Conclusion

The narrative of the Human Rights and Technology Conference that forms this chapter shows how there is an increasing awareness of human rights issues arising from digital technologies and that some parts of the Australian government are pushing back against unconscious adoption of digital technologies. This push-back does not occur in a vacuum but comes in response to rapidly transforming technologies and civil society efforts to show digital injustices. I read the AHRC's lead of a project on how human rights and technologies relate as evidence of the cut-through of civil society groups to call to account corporations and governments willing to rapidly implement untested and unreliable digital technologies. However, the Commission is a small group and has significantly less resources than other government entities.

The ways in which (digital) technologies were framed in the conference is a function of the polarising capacities of the more-than-real. There was a notable slippage throughout the day as digital technologies was reduced to AI: discussants talked about human rights in relation not to 'technology' broadly framed, and AI was the subfield of digital technologies of most interest and contention. The introductory and closing comments by Commissioners included this framing, as did the talk by the Chief Scientist of Australia. The more-than-real can obscure nuance and collapse differences in the production of grand narratives about the potential threats from digital technologies.

These critiques do not diminish the value of the Human Rights and Technology Issues Paper that the AHRC launched: it is an important step in developing a legal dialogue in Australian digital geographies that includes recognition of human agency. But it is not sufficient. More must be done to allow space for multiple agencies to find a place in this discussion space, for power relations to be reimagined for the more-than-human to be centred, and for sustainability (as problematic as it is) to have a seat at the negotiation table.

## References

ABC. (2018). *Technology—Friend or Foe for People with a Disability?* Big Ideas. Retrieved January 10, 2019, from https://www.abc.net.au/radionational/programs/bigideas/summer-technology/10209956.

Aira. (2018). *Aira: How It Works*. Retrieved April 5, 2019, from https://aira.io/how-it-works.

Australian Human Rights Commission. (2018). *Human Rights and Technology Issues Paper 2018*. Accessed August 10, 2018, from https://www.humanrights.gov.au/sites/default/files/document/publication/AHRC-Human-Rights-Tech-IP.pdf.

Bell, D. (2018). Food for Thought: How to Be Human in the Digital World. In *Human Rights and Technology Conference 2018, Sydney*. Retrieved June 5, 2019, from https://youtu.be/1nzvjQWr0kw.

Center for Humane Technology. (2018). *The Problem*. Retrieved October 20, 2018, from http://humanetech.com/problem.

Cisco. (n.d.). *Cisco Search on Google*. Retrieved October 26, 2018, from https://www.google.com.au/search?q=cisco&oq=cisco&aqs=chrome..69i57.1687j0j7&sourceid=chrome&ie=UTF-8.

Clarke, M. (2018). *Metadata Laws Under Fire as 'Authority Creep' Has More Agencies Accessing Your Information*. ABC News. Retrieved May 10, 2019, from https://www.abc.net.au/news/2018-10-19/authority-creep-has-more-agencies-accessing-your-metadata/10398348?smid=ABCscience-Twitter_Organic&WT.tsrc=Twitter_Organic&sf200495752=1.

Country, B., Suchet-Pearson, S., Wright, S., Lloyd, K., Tofa, M., Sweeney, J., et al. (2018). *Goŋ Gurtha: Enacting Response-Abilities as Situated Co-becoming*. Environment and Planning D: Society and Space. https://doi.org/10.1177/0263775818799749.

*Disability Discrimination Act 1992* (Cth).

Finkel, A. (2018). *Transcript of 'What Kind of Society Do We Want to Be?' Keynote for the Human Rights and Technology Conference*. Retrieved August 1, 2018, from https://www.chiefscientist.gov.au/wp-content/uploads/Human-Rights-and-Technology.pdf.

Galloway, K. (2017). Big Data: A Case Study of Disruption and Government Power. *Alternative Law Journal, 42*(2), 89–95. https://doi.org/10.1177/1037969X17710612.

*Human Rights Act Queensland 2019* (Queensland).

Iwarsson, S., & Ståhl, A. (2003). Accessibility, Usability and Universal Design—Positioning and Definition of Concepts Describing Person–Environment Relationships. *Disability and Rehabilitation, 25*(2), 57–66.

Kleeman, S. (2016). *Here Are the Microsoft Twitter Bot's Craziest Racist Rants*. Gizmodo. Retrieved October 22, 2018, from https://www.gizmodo.com.au/2016/03/here-are-the-microsoft-twitter-bots-craziest-racist-rants/.

Langton, M., & Mazel, O. (2008). Poverty in the Midst of Plenty: Aboriginal People, the 'Resource Curse' and Australia's Mining Boom. *Journal of Energy & Natural Resources Law, 26*(1), 31–65.

McLean, J., & Mackenzie, R. (2019). Digital Justice in Australian Visa Application Processes? *Alternative Law Journal*. https://doi.org/10.1177/1037969X19853685.

Neff, G., & Nagy, P. (2016). Automation, Algorithms, and Politics| Talking to Bots: Symbiotic Agency and the Case of Tay. *International Journal of Communication, 10,* 4915–4931.

Pash, C. (2018). *Sydney Is Becoming One of the Five Wealthiest Cities in the World*. Business Insider Australia. Retrieved October 6, 2018, from https://www.businessinsider.com.au/sydney-wealthiest-cities-2018-6.

Porter, L. (2010). *Unlearning the Colonial Cultures of Planning*. London: Ashgate.

*Racial Discrimination Act 1975* (Commonwealth).

*Sex Discrimination Act 1984* (Cth).

Sneddon, C., Howarth, R. B., & Norgaard, R. B. (2006). Sustainable Development in a Post-Brundtland World. *Ecological Economics, 57*(2), 253–268.

Tsoundarou, A. (2017, April). Technology Fail: The Centrelink Fiasco and the Impact on Vulnerable People. *Bulletin (Law Society of South Australia), 39*(3), 24–25. Retrieved July 20, 2018, from https://search-informit-com-au.simsrad.net.ocs.mq.edu.au/documentSummary;dn=800210027414593;res=IELHSS.

Van Dooren, T., & Rose, D. B. (2012). Storied-Places in a Multispecies City. *Humanimalia, 3*(2), 1–27.

Yu, D., & Hang, C. C. (2010). A Reflective Review of Disruptive Innovation Theory. *International Journal of Management Reviews, 12,* 435–452. https://doi.org/10.1111/j.1468-2370.2009.00272.x.

# 4

# Digital Rights and Digital Justice: Defining and Negotiating Shifting Human–Technology Relations

The digital is not inherently a space of oppression or emancipation, where capitalism is always unfettered and individuals' agency universally quashed. As Barassi (2015) offers, we can understand the impact of technological structures that digital capitalism produces while also appreciating that individuals and organisations challenge, contest and *remake* these structures through negotiated and everyday engagements. On each occasion that people open their Facebook or Twitter apps, they may not reflect on the corporate structures that enable these spaces in the digital, the worlding that these software produce. The everydayness of these actions can escape critical reflection of corporate structures and individual agency. After all, social media has ubiquitous qualities in certain contexts around the world and is an everyday, usual activity for many, including the surveillance of self and others that accompanies use of such public–private spaces.

But there are moments where digital injustices are accentuated. For example, Digital Rights Watch in Australia has been working on campaigns to protect Australian citizens from digital rights infringements. Lizzie O'Shea, legal scholar and Board member of Digital Rights Watch, described a campaign in 2018 to stop problematic federal laws


J. McLean, *Changing Digital Geographies*,
https://doi.org/10.1007/978-3-030-28307-0_4

surrounding encryption. The Assistance and Access Bill aimed to allow law enforcement agencies to require technology companies and others to build tools to circumvent encryption for law enforcement and intelligence gathering reasons. Digital Rights Watch commissioned Juice Media to make a video about this Bill to demonstrate its problems and invite submissions to point out the limitations of these encryption changes. The response to the video speaks to its success:

> In an age where we're told that people don't care about their privacy, that they buy the argument that if you've got nothing to hide then you've got nothing to worry about. In fact that's not true. 42,000 people took the time to issue – a standard format admittedly – but issue a submission to a draft exposure bill, which was much more than anybody else has managed to marshal in a long time. And also those people are prepared to give money. So then when we requested to appear at the Committee that the Bill is before, for example, we can go back to them and say, could you throw us a couple of bucks. Like you don't have to give huge amounts of money, but just to cover the costs, we're making a very specific ask. And they get it. Because it's following through on work that they've already helped us with and they appreciate it – the engagement and the investment that you put into that audience. (Lizzie O'Shea interview, November 2018)

The Senate Committee that heard the concerns of groups including Digital Rights Watch gave an audience to these issues of security and privacy yet ultimately the Bill passed Federal Houses of Parliament, with some amendments, as both major parties supported it. The galvanising of political support in this instance was not sufficient to stop the trumping of privacy concerns for surveillance priorities.

This chapter gives an overview to how digital rights are understood in Australia and considers if these are positioned in relationship to notions of environmental sustainability. More-than-real spaces can produce atomised framings of human–environmental connections and dissociate campaigns for digital rights from other struggles. By presenting perspectives from digital rights activists and narratives of social and cultural digital interventions, this chapter shows how environmental concerns are not always sidelined, and that more needs to be done to strengthen digital rights claims. Conversations about digital rights and environmental values show that the more-than-real is not a fixed, static space but full of possibilities and heterogeneities, similar to non-digital spaces (Massey 2005).

Digital rights are increasingly important as digital technologies are becoming deeply entrenched in everyday lives—sometimes without reflection on the costs and benefits of these affordances. Corporations and governments, civil society and academia, communities and nonhumans, all have roles to play in how these digital rights are understood and protected. The way we conceive of digital rights is important as what is deemed part of the digital rights conversation impacts upon what rights are maintained or sidelined. In digital geographic literature, digital rights are framed as relevant to development issues (Kleine 2019, 234) in terms of access to digital tools and also with respect to 'data protection, privacy, and freedom of expression' and in contexts of geopolitical power negotiations and data control (Crampton 2019). An example of such geopolitical wrangling is the concern that Japan, Australia and New Zealand hold about Chinese company Huawei's involvement in internet expansion (through 5G systems) and the Chinese government's potential to exploit such technological control for its own purposes (Inkster 2019). Digital rights are also defined as spanning across individual freedoms to the capacity to complete everyday activities and are connected to rights already conceptualised in non-digital contexts.

## A Narrative of a Digital Campaign Against Government AI Fail: #notmydebt

Digital rights campaigns are supporting individuals to counter infringements by corporations and governments. A powerful example of a recent and successful digital rights campaign arose in response to Australian federal government overreach with automation technology. Since mid-2016 the federal government has automatically generated debts for social security recipients by data matching with the Australian Tax Office (ATO). Centrelink is the organisation that administers social security payments from the aged pension to unemployment payments.

The problem with this data matching is that it involved cross-checking fortnightly income data from Centrelink with annual income information from the ATO. Incorrect debts and general confusion ensued as the data matching was fundamentally flawed. Kate Galloway outlines one scenario for false debt generation:

> if a person was unemployed for six months and receiving benefits, but employed for the second six months in a high paying job, the six months of income will show up as averaged payments fortnightly for the entire year. That person will be asked to prove their Centrelink entitlement by producing pay slips for the period of employment. If unable to do so, Centrelink will claim repayment. (Galloway 2017, 93)

The robodebts included high numbers of miscalculations: Tsoundarou (2017) reported that of the 20,000 debts automatically generated by the Department of Human Services, 4000 were false. Many wrongly imputed debtors did not challenge the debts and simply paid Centrelink to avoid further problems. If people did appeal the decisions, they had to go through long appeal processes to get these debts cleared, adding an emotional and intellectual cost to a process they should not have been party to in the first place.

Given these challenges, civil society organisations and activists responded with strategic digital campaigning. Lyndsey Jackson began a digitally-mediated campaign under the hashtag #notmydebt and its success led to her current role as Chair at EFA. The stories shared under the hashtag were then collected and recorded on the Not My Debt website to show the extent of the problem (Jackson, interview 2018). Over 700 stories of people who were forced to start payments while still appealing decisions shared their narratives of government ineptitude. The campaign ultimately led to a federal government Senate Inquiry into how and why the false debts were decided. The website home for the #notmydebt campaign provided people with clear instructions for how to create a good submission for the Inquiry, including a 5 step overview of what to do. The benefit of this approach was that it didn't duplicate digital processes as it involved 'using digital tools to support people to better engage with the systems that are already in place, rather than using new digital tools to create a bulk generated mail app which only carries so much impact in government' (Lyndsey Jackson, interview October 2018). Helping people participate in formal governance processes, such as a Senate Inquiry, is seen as an effective part of digital campaigning.

Analysis of the #notmydebt campaign from a social media perspective by Bruns et al. (2017) demonstrates that the digital rights activists effectively framed the robodebt Centrelink process as a form of administrative malpractice. Bruns et al. (2017) consider over 400,000 tweets and found that Asher Wolf, a digital activist, was central to much of the campaigning on Twitter. During the peak of the #notmydebt campaign, George Brandis, Attorney General for Australia, appeared on the national broadcaster's flagship political debate panel programme Q&A,[1] defending his government's handling of the matter, but this did not make a substantial impact on the campaign (Bruns et al. 2017). Brandis suggested that it was easy to remedy any false debts: people just had to contact Centrelink to explain the situation (ABC News 2017). The audience was quick to respond with laughter as it is well known how long it takes to get through to Centrelink and many calls do not get through at all; in 2017 alone, more than 33 million calls were not even answered by the government agency (Dingwall 2018). Brandis' input did little to stem the tide of dissatisfaction with Centrelink and the Senate Inquiry proceeded. The report from the Inquiry was published in June 2017 and recommended a complete overhaul of the automated decision-making to make sure that procedural fairness would be a part of the debt-raising process in the future, and a review of all debts that had been raised through the system. However, in early 2018, further complaints and calls for a fresh inquiry were raised as issues as debt-raising continues and the legitimacy of new debts is under dispute. Centrelink has cancelled 40,000 robodebts as of February 2019 and, at time of writing, legal action is underway in some parts of Australia to further appeal the way the algorithm works (Henriques-Gomes 2019).

---

[1]Q&A itself has a strong social media presence as it uses a 'livefeed' of Twitter commentary from people not in the audience as a banner during its broadcast.

Using big data and automated systems to streamline governance can potentially increase efficiencies and reduce burdensome, repetitive tasks for some employees. However, if governments tend to use data matching tools without recognising basic procedural fairness and the impacts of errors in systems, then the governance processes that rely on such automation are inherently unjust. Galloway (2017) argues that governments should not blindly apply automated processes, such as Centrelink's robodebt approach, as citizens bear the cost of errors—and these can be substantial. Some people on the #notmydebt website list debts of five thousand Australian dollars (Not My Debt 2017), and the stress and emotional cost of these claims is important, if less quantifiable than the debts themselves. The social activism that challenged these erroneous debts was powerful. This example of the constructive possibilities of digital spaces, or what Daly et al. (2019) call 'good data', is compelling. Good data stories like this one—challenging the 'bad' data that government has produced—may contribute to a more 'fair and just digital economy and society' (Mann et al. 2019, 8).

## Thinking on Digital Justice

Digital justice is an emerging concept that refers to the way digital technologies may enable or constrain pursuit of just outcomes in multiple contexts. In legal literature, Fileborn writes about digital justice in an analysis of Hollaback Melbourne, a site hosted by Tumblr, which aims to provide a safe space for women to share stories of sexual harassment. Hollaback! Melbourne is a good start at working towards digital justice for women, Fileborn argues, but ultimately is somewhat insufficient in facilitating justice for sexual harassment survivors. Digital justice is not explicitly defined in Fileborn's article and her conclusion suggests that important gaps remain in obtaining justice for sexual harassment survivors despite efforts such as Hollaback! in the digital.

For the purposes of this book, digital justice is defined broadly as attempts to achieve just outcomes and processes through digital interventions and in digital spaces, from social media campaigns, crowdfunding initiatives, ad hoc digital collectives, to sharing strategies and ideas in digital spaces. Digital injustices emerge when digital processes and spaces are not inclusive in terms of recognising and including diverse stakeholders, and ultimately do not produce equitable distribution of the costs and benefits of those processes. I draw on David Schlosberg's (2004) reframing of environmental justice to develop this digital justice concept here, focusing on three key notions of justice, inspired by global environmental justice movements. First, Schlosberg (2013) argues that there needs to be equity in the distribution of environmental risk, then recognition of the diversity of participants and experiences in communities, and last, participation in processes relating to environmental decision-making.

If, for example, we only focus on how the digital enables the voice of often marginalised people—in Schlosberg's conceptualisation this would speak to the *participation* circle of concern—this may not be sufficient in achieving a just process and outcome. Couldry et al. (2013) discuss crowded digital spaces where many voices are possible but only some are heard and contextualising this issue in the digital justice frame means that participation is not necessarily sufficient. Rather, recognition of diverse actors and distribution of benefits (cf. Rawls' (1972) social justice theory) are also required. Environmental justice practice and theories began with an analysis of racially differentiated distribution of environmental costs in northern American contexts and has expanded, horizontally and vertically, to encompass a wide range of issues, from food security to climate justice (Schlosberg 2013).

Applying this theory to digital justice, there must be equity in terms of the distribution of digital access, on all levels, next there should be recognition of the diversity of voices and experiences of people relying on digital technologies and, third, participation in processes that affect people and communities using digital spaces. Digital injustices produce and exacerbate digital divides, where uneven access to technologies is socio-economically, culturally, and spatially constructed. We need to address digital inequalities (Robinson et al. 2015), in all their forms, rather than perpetuate and/or produce more digital injustices through transforming social systems and digital spaces.

As well as legal literature on digital justice, academic work in this area includes media and communications scholars. For example, Dencik et al. (2016) define 'data justice' as analysing 'the implications that data-driven processes at the core of surveillance capitalism have for the pursuit of substantive social and economic justice claims'. Dencik et al. draw on insights from activists engaged with revealing and resisting surveillance in the United Kingdom, and note that a consumer group in the United States used 'Data Justice' as a conceptual framing to approach big data as an economic justice concern. Dencik is co-founder/leader of the Data Justice Lab in the United Kingdom, bringing together critical scholarship on the affordances of digital technologies. Digital justice in this chapter builds on data justice approaches by adding to the data justice focus on data-driven processes in surveillance to look at changes relating to the digital more broadly, including multi-scaled justice questions that emerge when hegemonic systems are challenged.

The relatively small and fleeting gesture of some digital protest can seem insignificant compared to more sustained, offline efforts. However, these digital gestures enable even the smallest surges of emotion and affect to generate something, and then cumulatively, individual minor acts coalesce to form moments and possibly movements in digital affect cultures (Döveling et al. 2018) that stem from, but are not limited to, the more-than-real. The scope and possibilities arising from digital justice efforts are broad and are common to efforts to pursue digital justice is a capacity for passing desires, in a Deleuzean sense, to be transformed into substantive interventions, where desire is not about lack but a positive force (Gao 2013).

We can see interactions between the mundane and the spectacular delivering unexpected outcomes in the digital, demonstrating the unpredictability of the more-than-real. The visually arresting fundraising for Amyotrophic Lateral Sclerosis (ALS) facilitated by the Ice Bucket Challenge that began in 2014 (Sohn 2017) is a recent example of the unexpected in the more-than-real. The Ice Bucket Challenge involved filming of the pouring of a bucket of ice and water over a participant's head and daring others to follow suit by sharing YouTube videos. It was immensely popular and celebrities and everyday people took part: more than 17 million people engaged by uploading videos that were watched billions of times around the world (Sohn 2017). But it was not just the spectacle that succeeded as funding resources for ALS researchers and institutions grew from donations—by 2015 approximately 28 million people had donated an estimated $220 million worldwide to the ALS cause (Sohn 2017). The Ice Bucket Challenge was amusing and visceral to watch and yet there was nothing intrinsic about its initiative that necessarily suggested it would achieve such reach and success. It is the accretion of smaller somethings that might be at play here (Stewart 2007). More-than-real geographies do cultural, political, economic and social work with a blend of emotion, affect and action—sometimes working towards socially just outcomes, and at other times not.

## Spontaneous Digital Justice in Action

Individuals, groups and institutions are aiming to use digital justice avenues to build more equitable societies with broad agendas, addressing injustices that pre-exist the expansion of digital geographies. For example, Indigenous people are using social media for different purposes, from the prosaic to the transformative (Duarte 2017), strategically drawing on digital tools to intervene in political matters when and where possible. Duarte (2017, 10) rightly points out that 'the expression of strong mediatised Indigenous voices does not ensure that dominant government authorities will accordingly listen or act in a just manner'; significant structural changes are required for that to happen, including decolonising of spaces, places and institutions. Chapter 5 discusses decolonising digital spaces in more detail but here I will focus on a digital justice campaign that was driven by Indigenous people as an instance of spontaneous digital justice.

There is no singular way to seek digital justice; multiple opportunities to push for more just outcomes through digital actions are captured in research on ad hoc publics (Bruns and Burgess 2011) and issue publics (Kim 2012). The ongoing struggles for Indigenous peoples to seek social justice is a focus for Aboriginal scholar Bronwyn Carlson who has been examining Indigenous people's activism within digital spaces in specific ways. For example, Carlson and Frazer (2016) describe the #SOSBLAKAUSTRALIA campaign, started in Wangkatjungka by the Bieundurry family, that aimed to draw attention to the state government closure of remote Indigenous communities in Western Australia. #SOSBLAKAUSTRALIA was intended to challenge the political decision-making that was marginalising Indigenous people from formal governance processes, and their country, once again, in another wave of setter colonial incursion (Carlson and Frazer 2016). In 2014, the Western Australian state government was discussing closing up to 150 of 274 remote Indigenous communities for not being economically viable (Howitt and McLean 2015). The #SOSBLAKAUSTRALIA campaign aimed to alert people to the effect of these processes and assert the rights of Indigenous people's occupation on Country, in remote Australia. Indigenous rights movements such as this example of a social media movement reinforces, rather than replaces, Indigenous cultural practices and relations (Carlson and Frazer 2016). The technology does not, in and of itself, enable the social and political change required for decolonisation but it may help facilitate and support the success of offline movements towards that end. Carlson et al. (2017, 1) introduce the term 'shared recognition' that captures how Indigenous people experience a 'collective sense of anger and frustration experienced by Indigenous people when traumatic events in the public domain act as reminders of the enduring-ness of colonial violence' and then act on social media to resist this violence.

The power of #SOSBLAKAUSTRALIA comes from the digital facilitating Indigenous people's sharing of similar experiences across distance, especially useful for widely dispersed populations such as those in remote Indigenous communities of Western Australia. Anger accompanied the circulation of #SOSBLAKAUSTRALIA—anger at another governance intervention that would further reduce access to Country was captured in the digital movement and circulated in offline actions. Like other digital activist efforts, #SOSBLAKAUSTRALIA used Twitter and Facebook pages to build momentum and within three weeks of the #SOSBLAKAUSTRALIA Facebook launching, 50,000 likes were recorded on the page (Carlson and Frazer 2016) and as of December 2018, over 68,000 likes are recorded on that page, as shown in Fig. 4.1. The campaign to stop the forced closure of Indigenous communities included more traditional protests such as rallies and occupying public spaces alongside, and facilitated by, #SOSBLAKAUSTRALIA's

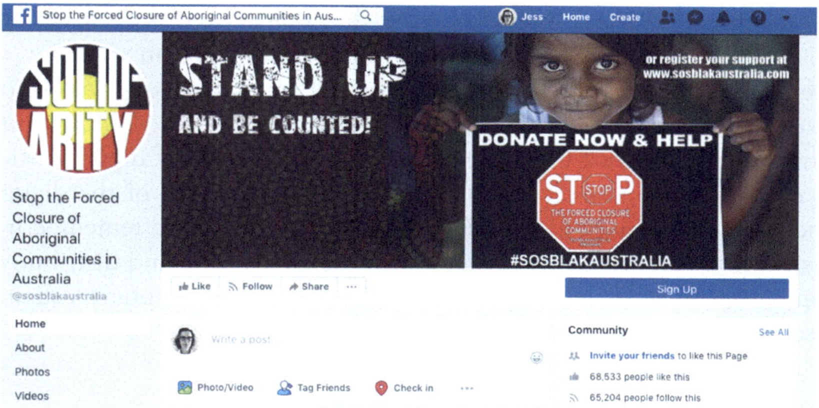


**Fig. 4.1** Screenshot of 'Stop the Forced Closure of Aboriginal Communities in Australia @sosblakaustralia' Facebook page

social media presence (di Stefano and Clarke 2015) and the digital and non-digital actions all made affect visible. Carlson and Frazer (2016) are careful to not overstate the role of digital activism in contributing to substantial changes for Indigenous peoples, especially as the outcomes of the actions relating to #SOSBLAKAUSTRALIA remain uncertain. However, Carlson and Frazer (2016) emphasise that the digital has helped to facilitate citizenship action for Indigenous people who previously struggled to get a prominent voice.

Media theorists have considered aspects of digital justice and I draw on some of that work here for its explicit relevance to digital geographies. In a dialogue-based article that examines digital in/justice, Couldry et al. (2013, 611) argue that 'digital media don't automatically "solve" issues of voice' suggesting that simply because the means of communicating positions and ideas is available, this opportunity may not translate into widespread recognition of that voice. The unpredictability of social movements that use digital technologies speaks to this attention issue. Why do some hashtags, blogs and videos succeed on their terms while others fail? Why are some digital gestures corrosive while others are productive? Katsh and Rabinovich-Einy (2017) argue that the digital can produce disputes and social problems but can also provide solutions for contentious issues. 'Digital Justice: Technology and the Internet of Disputes' (Katsh and Rabinovich-Einy 2017) provides an overview of their legal approach to digital problems, with an emphasis on procedural justice. While recognizing the value of these approaches, I draw upon environmental justice literature to develop the notion of digital justice to emphasise the spatial, emotional and affectual aspects of the digital. The more-than-real begs analysis of procedural justice and, remembering the work of Sedgwick's *besides* approach, the recognition and distributive approaches in all their forms as outlined so far in this chapter on digital justice.

## A Snapshot of Digital Rights in Australia

A report on 'Digital Rights in Australia' (Goggin et al. 2017) discusses perspectives on digital rights in terms of privacy and data analytics, government data matching and surveillance, work-based digital rights and freedom of speech in the digital. The compelling report draws on a survey of 1600 Australians and focus groups, offering an analysis of legal, policy and governance issues relating to digital rights. The report defines digital rights in a broad sense:

> It explores the ethical and legal challenges we face in using digital, networked technologies and the debates we are having about how to best manage their transformative impacts. (Goggin et al. 2017, 1)

Gender was an important differentiating factor in perspectives on privacy and safety in digital spaces, and regulation of social media especially. For example:

> Women experience the online world differently from men: they are more likely to agree that they actively protect their privacy online (71%, compared with 63% of men) and change their social media settings (63%, compared with 58% of men), but feel no more in control of their privacy (39%, compared with 38% of men). (Goggin et al. 2017, 1)

The report does not explicitly discuss non-binary gender issues but does stipulate the importance of recognizing different digital rights for different actors, including in terms of sexuality, age and people with disabilities. The survey respondents were asked to identify gender as Male/Female/other (Goggin et al. 2017, 62) and in all of the tables offering data over the 66 page report, just Male/Female categories are listed. Perhaps no one who identified as non-binary gendered responded to the survey. The binary construction of gender in the report is evident in statements such as 'both genders reporting a similar degree of unwanted sexual contact' (Goggin et al. 2017, 44). Such binary constructions may serve to perpetuate heteronormativity and limit recognition and inclusion of diverse groups and individuals. The risks of ignoring these sorts of difference include that some positionings are already marginalised when defining digital rights, and space for queering digital technologies (Gieseking 2017) is reduced.

Digital rights relate to freedom from abuse and violence, as well as freedom of speech. Gendered differences in the assertion of these digital rights are captured in Goggin et al. (2017)'s report:

> Men were less likely than women to agree with the need to remove within 24 hours instances of sexual harassment, abuse targeted at an individual, or hate speech that encourages violence against others. Women were less supportive than men of the right to anonymity. (Goggin et al. 2017, 2–3)

The fact that women responded to the right to anonymity differently to men points to the different lived experiences of digital spaces (Bates 2014; Ferreira and Salvador 2015). Research from a range of fields corroborates the anonymity problem with digital spaces and experiences of gendered sexual harassment, including work on digital gaming (Ruvalcaba et al. 2018) and journalism (Chen et al. 2018). Anonymity in digital spaces can enable damaging behaviours as geographies of responsibility are warped.

The Digital Rights report gives nine policy recommendations directed to Australian governments and transnational corporations. Overall, the people who participated in the study do not think that Australian governments and corporations are doing enough to protect privacy and that the Federal Government itself is not protecting digital rights. Better regulation and greater recognition of individual rights is a common theme to many of the recommendations. O'Shea (2019) shares this view and argues that greater regulation is required for digital technologies to help build more democratic futures. People are willing to negotiate some personal interest for the sake of greater collective security '[b]ut current policy moves to collect and centralise more data—through My Health Record or a Digital ID programme—look like pushing beyond what Australians are comfortable with' (Goggin et al. 2017, 3). It is difficult to imagine that the Australian government would proactively seek to protect digital rights when it is simultaneously introducing more measures to challenge the privacy of its citizens. Law reform and better education about media law were key recommendations of the report as well.

Alongside this snapshot of how digital rights are understood in Australia, digital rights are being defined in relation to the pre-existing human rights context. The Electronic Frontiers Australia (EFA) not-for-profit group describes themselves as 'promoting and protecting digital rights in Australia since 1994' (EFA, n.d.). Lyndsey Jackson, Chair of EFA, offered her insights into what digital rights mean:

> Look it's pretty broad because the starting point that we talk about, and particularly, with respect to the Assistance and Access legislation,[2] is that Australia has no Bill of Rights, so because we don't have a starting point of human rights and how we prioritise and preserve human rights, then that makes any digital rights legislation or issues problematic because we're not protecting human rights first and foremost. So even though that's a non-technical concept, that's ingrained within the approach to digital rights. And then really it's anything where we've got Internet technologies and other digital technologies that impinge on the rights of people. So, that can be things like surveillance, privacy, access to Internet, access to technology, those sorts of things. (Interview with Lyndsey Jackson, October 2018)

Here, digital rights refer to digital technologies that impinge on the rights of people as well as the capacity to access digital technologies.

[2]The Assistance and Access Bill was introduced to Australian parliament in September 2018.

Rights can be obstructed by digital technologies rather than just facilitated through these tools in this framing. But the broader context of the lack of a Bill of Rights in Australia is crucial in this understanding of digital rights. Within Australia, human rights are protected through a range of pieces of Federal Legislation, while at the state level some protections have been enshrined. Queensland has now passed a Human Rights Act that enshrines legislative protection of a range of human rights and this follows the Victorian and Australian Capital Territory's lead on enacting human rights legislation. Groups like the Australian Privacy Foundation (APF) and the Queensland Council for Civil Liberties (QCCL) campaigned for the introduction of the Human Rights Act to secure protection of 23 human rights, including the following that are directly related to digital rights:

- Recognition and equality before the law
- Right to life
- Protection from torture and cruel, inhuman or degrading treatment
- Freedom from forced work
- Freedom of thought, conscience, religion and belief
- Peaceful assembly and freedom of association. (Bavas 2018)

The *Human Rights Act* 2019 of Queensland includes reference to technology when outlining rights for people in criminal proceedings as 'to have the free assistance of specialised communication tools and technology, and assistants, if the person has communication or speech difficulties that require the assistance'.

Protection of digital rights has also emerged as a priority for not-for-profit organisations that are focused on safeguarding privacy. For instance, the APF has been responding to threats to digital rights pertaining to privacy over the past five years but began as an organisation that was advocating against privacy incursions more generally. From an interview with Mark Walkom, secretary of the APF, it was clear that the APF have shifted over time to respond to changing digital technologies. The APF began as a grassroots organisation of people concerned with the possibility of an identity card—the 'Australia Card'—and activists used traditional protest modes to mobilise the population to resist its introduction.

As the organisation grew, the APF board, comprised mostly of academics and lawyers, tended to work together to respond to proposed legislation or Discussion Papers and formulated considered responses. Now, another shift is happening as collaborations with EFA and other organisations to formulate social media driven campaigns, such as that against My Health Records campaign in 2018. At the same time, digital technologies are central to the work that the organisation does:

> Well, I guess we're predominantly digital, so we're distributed around Australia. We have a bit of a clustering in Sydney and Brisbane. A lot of the work that we do is digitally based, so whether that's on various chat platforms, whether that's on our website, by the traditional sort of mailing list type things, 99% of what we do is probably based online. (Interview with Mark Walkom, October 2018).

Echoes of Pickerill's (2003) observations in *Cyberprotest* that digital technologies are enabling effective and more powerful activism are clear in the APF's strategic approach. Although Pickerill (2003) was discussing environmental activists in Britain, similar modes are clear for privacy activists over fifteen years later and in another part of the world, as the organisations that she describes in that text changed in response to digital technologies, just as APF and EFA are changing. Also noteworthy is the flexibility and responsiveness of EFA and APF to the circumstances, opportunities and limitations of their advocacy. Pragmatically drawing on digital technologies to enable and expand advocacy is a common theme of the EFA, APF and other advocacy groups that are aiming to protect digital rights (Brady et al. 2015).

Digital justice is being pursued in a range of contexts, perhaps most visibly on a global scale in movements like #MeToo (Rodino-Colocino 2018). Feminist digital activism is working to draw attention to sexist and misogynistic behaviour in diverse ways and on a range of issues (among many others see McLean and Maalsen 2013; Oberhauser et al. 2017). Powell and Henry (2017) develop a legalistic analysis of digital justice with respect to sexual violence in multiple contexts, such as the hashtag that highlighted the abuse and kidnapping of Nigerian girls #bringbackourgirls and #yesallwomen that gathered critiques of everyday sexism and its ubiquity. The following quote shows how they frame digital justice:

> ...hashtag activism contributes to the development and proliferation of feminist counter-discourses of sexual violence, its nature and causes, sometimes functioning as a safe space for victim-survivors and their allies, sometimes extending to challenge white, middle-class, heterosexual male dominance in the public sphere. In doing so, hashtag activism can be understood as a technosocial political strategy directed at challenging the gender misrecognition that underlies sexual violence more generally. (Powell and Henry 2017, 285)

Powell and Henry (2017) distinguish between digital technologies helping people achieve 'digital' or 'informal' justice, and online feminist activism. This distinction may not be particularly useful in light of the messiness that the more-than-real facilitates. Following the lines of argument offered in this chapter, online feminist activism can be viewed as within the digital justice remit—focusing on gendered inequities, gender-based violence, gender identity and more.

## Environmental Sustainability and Digital Rights Action

Conversations about environmental sustainability and digital rights action do intersect on occasion but not as frequently as needed in this Anthropocene moment, and this may be due to older divisions between human rights and environmental sustainability thinking. Woods (2007) provides interesting insights into the way human rights and environmental sustainability share features in that proponents of both fundamentally express a concern for others, ecological or social. In the context of globalisation, Woods (2007) argues that human rights and green activists and theorists point to global forces of change as producing, and deepening, inequalities and environmental degradation.

Digital spaces are multiply scaled and are often globally networked. A paradox could be seen with respect to the relationships between human rights and environmental sustainability and priorities relating to the achievement of both. Human rights can be framed as immediate and pressing concerns while environmental sustainability is a long-term objective since environmental risks are not always close to hand. For instance, the planetary systems that are currently under threat, including climate and biodiversity, are sometimes held at a distance in deeply urban lives (a false consciousness challenged by critical geographic scholars including Reddy (2018)). Therefore, human rights attainment may be of higher concern to those engaging with questions of justice, than environmental sustainability. Human rights are, however, linked to a liberal democratic system that is at times hostile to environmental sustainability and therefore moving to the next area of justice within that social structure is problematic. There are some assumptions underlying this argument that need unpacking: first, that human rights are not actually strongly tied to environmental sustainability, and second, that social, cultural and political transformations can not achieve both goals simultaneously rather than consecutively.

From speaking with people working for the protection of digital rights in various contexts, there was a general sense that environmental sustainability is not always at the forefront of their minds but was a part of the mix. For instance, Mark from the APF gave the following response to a question about sustainability and digital rights activism:

> I mean, I think the physical environmental impact is pretty minimal, which is… it's always a positive thing. I think, interestingly, there's a social environment. It seems to be becoming a lot more polarised. You either are or you aren't. It's very black and white. There's a lot of loss of nuance and complexity. If you look at politics, it's become very sort of them and us. So I think that people either get drawn or forced onto these sides, so I think there's the social one. It's making it… It's easy to build up a bit of an echo chamber, but it's also increasingly difficult to get people that don't necessarily agree with you to have a reasonable discussion or think about things from a different angle. (Mark Waldom interview, October 2018)

The polarised quality of digital spaces is an interesting point in terms of the social sustainability of activism that relies on this space.

If nuance and subtlety is of limited use in digital spaces, then to rise above the cacophony and gain attention requires pushing to discursive edges rather than working on an integrated middle. Sunstein (2018) discusses the rise of echo chambers and how divisive social media has become in her analysis of democratic challenges in the United States. Social media users choose who to follow and who to block, where to gain information and where to ignore, creating echo chambers. For people working to achieve protection of privacy, such as Mark with APF, these echo chambers can be a substantial barrier to communicating and building campaigns. The pool of possible people to 'activate' with their message is reduced because of the polar tendencies of the more-than-real.

While environmental concerns may not be a top priority for all digital rights activists, it does emerge as an interest in certain contexts. For instance, Lyndsey Jackson, EFA's Chair, said that emergent dialogues about environmental sustainability and rights protection should be increasing rather than left until some distant other time.

> So I think that those discussions are lacking and things around either electricity use or devices and the number of devices that are out there and what that impact means and what it means when you have to get new devices. I think that that environmental cost is just so rarely, how you even get rid of metals and stuff like that, that are rare.
>
> I think that technologists are often quite progressive, but there can be a lot of travel, for example, that is involved within the internet communities and people don't think much. I got an email from Google today, asking me to, you know, 'do you want to go to Paris?' It's like – three days in Paris. On the one hand, that seems like such an opportunity, on the other hand to do that for a few days just, I just think of the carbon miles and it actually does seem quite wasteful. (Lyndsey Jackson interview, October 2018)

The costs and benefits analysis that emerges in this reflection on a possible networking opportunity for EFA is of particular relevance to the mechanisms of the more-than-real. Google would have the best technology available to run such a meeting as a digital event—the lengthy and carbon-intensive flight to Paris could easily be avoided. But the materiality of an embodied presence is valued greatly, even for an institution that is one of the biggest tech companies in the world, and one that has a contentious record of giving into governments over issues such as censorship. Google's foray into Chinese digital spaces in 2018 and its concession to censoring sites that were critical of the Chinese government is a notable example of this, with suggestions it may be building a censor-friendly Google search engine (Yuan and Wakayabashi 2018). The targets of digital rights campaigns tend to shift, from corporate to government bodies, individuals to communities, depending on the issues at play and the resources at hand.

## Conclusion

Defining digital rights involves defending the human rights of people in digital spaces, as well as preventing the infringement of human rights by digital technologies. The #notmydebt campaign is an excellent example of a moment where digital rights were compromised by an automated process *and* defence of these was facilitated by digital spaces, namely Twitter and a website. But these relatively novel forms of activism brought about a more traditional mechanism of change in the shape of a Federal Government Senate Inquiry. The way the more-than-real works is through these occasionally contradictory processes of incursion and resistance. The power of governments to pursue the adoption of automation, for example, for efficiency purposes and to increase revenue, is clearly appealing. But the consequences of doing so to welfare recipients that may already have challenging circumstances, be that through unemployment or sole parenting responsibilities among many others, seems particularly unjust.

In this chapter, I have brought together discussions of what digital rights are with issues relating to environmental sustainability, keeping in mind the debates circulating with the use of these terms. The question of what is reasonable and appropriate in an environmental sustainability sense for digital technologies is sometimes, but far from always, at the forefront of activists' minds that are working in the digital rights area.

The perceived immateriality of the more-than-real may play a role in this, as may long histories of stratification of rights-based and environmental activism. Lawyers and legal academics tend to spend more time prosecuting and developing theories of human rights while environmentalists and environmental scientists agitate for environmental activism. Despite this, climate and environmental justice thinking is bringing together these debates and asking for simultaneous consideration of environmental and human rights. If we perceive the ongoing expansion of digital technologies in our everyday lives as having space for conversations around digital rights, then there is also space for bringing environmental sustainability along for that troublesome ride. Those who are advocating for protection and consideration of digital rights tend to focus on particular campaigns at opportune times—strategically calling for interventions and redress when necessary. The targets of these campaigns are corporations and governments, and sometimes these institutions are the same ones that are charged with human service-oriented work. Last, I argue that digital justice, perhaps necessarily, is a nebulous concept and practice, as the more-than-real helps to produce and perpetuate such messiness. Connecting digital justice debates to those on multispecies rights (Van Dooren et al. 2016) in future work may create further valuable insights into how agency and power work in the more-than-real.

## References

ABC News. (2017, January 21). *Q&A: Laughter as George Brandis Says Debt Recovery Notices Can Be fixed by Contacting Centrelink*. ABC News. Retrieved October 15, 2018, from https://www.abc.net.au/news/2017-02-21/q&a-george-brandis-says-centrelink-call-solves-welfare-disputes/8287482.

Barassi, V. (2015). *Activism on the Web: Everyday Struggles Against Digital Capitalism*. London: Routledge.

Bates, L. (2014). *Everyday Sexism*. London: Simon and Schuster.

Bavas, J. (2018). *Queenslanders will soon be Protected under a Human Rights Act*. Here's What that means for you ABC News, 31 October 2018. Retrieved May 10, 2019, from https://www.abc.net.au/news/2018-10-31/human-rights-act-queensland-parliament-brisbane/10448908.

Brady, S. R., Young, J. A., & McLeod, D. A. (2015). Utilizing Digital Advocacy in Community Organizing: Lessons Learned from Organizing in Virtual Spaces to Promote Worker Rights and Economic Justice. *Journal of Community Practice, 23*(2), 255–273.

Bruns, A., & Burgess, J. (2011). The Use of Twitter Hashtags in the Formation of Ad hoc Publics. In *Proceedings of the 6th European Consortium for Political Research (ECPR) General Conference 2011, University of Iceland, Reykjavik.* Retrieved July 20, 2018, from https://eprints.qut.edu.au/46515/.

Bruns, A., Moon, B., & Dhegan, E. (2017). *Dynamics of a Scandal: The Centrelink Robodebt Affair on Twitter.* Presentation at the Australian and New Zealand Communication Association Conference, University of Sydney. Retrieved October 10, from http://snurb.info/node/2244.

Carlson, B., & Frazer, R. (2016). Indigenous Activism and Social Media: A Global Response to #SOSBLAKAUSTRALIA. In A. McCosker, S. Vivienne, & A. Johns (Eds.), *Negotiating Digital Citizenship: Control, Contest and Culture* (pp. 115–130). London and New York: Rowman & Littlefield International.

Carlson, B., Jones, L., Harris, M., Quezada, N., & Frazer, R. (2017). Trauma, Shared Recognition and Indigenous Resistance on Social Media. *Australasian Journal of Information Systems, 21.* http://dx.doi.org/10.3127/ajis.v21i0.1570.

Chen, G. M., Pain, P., Chen, V. Y., Mekelburg, M., Springer, N., & Troger, F. (2018). 'You Really Have to Have a Thick Skin': A Cross-Cultural Perspective on How Online Harassment Influences Female Journalists. *Journalism.* https://doi.org/10.1177/1464884918768500.

Couldry, N., Gray, M. L., & Gillespie, T. (2013). Culture Digitally: Digital In/Justice. *Journal of Broadcasting & Electronic Media, 57*(4), 608–617. https://doi.org/10.1080/08838151.2013.846343.

Crampton, J. (2019). Geopolitics. In J. Ash, R. Kitchin, & A. Leszczynski (Eds.), *Digital Geographies* (pp. 281–290). London: Sage.

Daly, A., Devitt, K., & Mann, M. (2019). *Good Data.* Brisbane/Meanjin: Queensland University of Technology. Retrieved June 19, 2019, from http://networkcultures.org/wp-content/uploads/2019/01/Good_Data.pdf.

Dencik, L., Hintz, A., & Cable, J. (2016) Towards Data Justice? The Ambiguity of Anti-surveillance Resistance in Political Activism. *Big Data & Society*, 1–12. https://doi.org/10.1177/2053951716679678.

Dingwall, D. (2018). More Than 33 Million Centrelink Calls Unanswered as DHS Denies Staffing Problem. *Sydney Morning Herald.* Retrieved February 1, 2018, from https://www.smh.com.au/public-service/more-than-33-million-centrelink-calls-unanswered-as-dhs-denies-staffing-problem-20180307-h0x4up.html.

di Stefano, M., & Clarke, A. (2015). *Indigenous Rights Protestors Shut Down Sydney and Melbourne.* Buzzfeed. Retrieved April 1, 2019, from https://www.buzzfeed.com/markdistefano/aboriginal-protests-shut-down-cities.

Döveling, K., Harju, A. A., & Sommer, D. (2018). From Mediatized Emotion to Digital Affect Cultures: New Technologies and Global Flows of Emotion. *Social Media & Society, 4*(1). https://doi.org/10.1177/2056305117743141.

Duarte, M. (2017). Connected Activism: Indigenous Uses of Social Media for Shaping Political Change. *Australasian Journal of Information Systems, 21.* Retrieved March 1, 2019, from http://marisaduarte.net/Duarte_Connected%20Activism.pdf.

EFA. (n.d.). *EFA Twitter Profile.* Retrieved August 4, 2019, from https://twitter.com/efa_oz.

Ferreira, E., & Salvador, R. (2015). Lesbian Collaborative Web Mapping: Disrupting Heteronormativity in Portugal. *Gender, Place & Culture, 22*(7), 954–970.

Galloway, K. (2017). Big Data: A Case Study of Disruption and Government Power. *Alternative Law Journal, 42*(2), 89–95. https://doi.org/10.1177/1037969X17710612.

Gao, J. (2013). Deleuze's Conception of Desire. *Deleuze Studies, 7*(3), 406–420.

Gieseking, J. J. (2017). Messing with the Attractiveness Algorithm: A Response to Queering Code/Space. *Gender, Place & Culture, 24*(11), 1659–1665. https://doi.org/10.1080/0966369X.2017.1379955.

Goggin, G., Vromen, A., Weatherall, K., Martin, F., Webb, A., Sunman, L., et al. (2017). *Digital Rights in Australia.* Retrieved August 5, 2019, from http://hdl.handle.net/2123/17587.

Henriques-Gomes, L. (2019). Centrelink Cancels 40,000 Robodebts, New Figures Reveal. *The Guardian.* Retrieved May 10, 2019, from https://www.theguardian.com/australia-news/2019/feb/06/robodebt-faces-landmark-legal-challenge-over-crude-income-calculations.

Howitt, R., & McLean, J. (2015). Towards Closure? Coexistence, Remoteness and Righteousness in Indigenous Policy in Australia. *Australian Geographer, 46*(2), 137–145. https://doi.org/10.1080/00049182.2015.1020992.

*Human Rights Act 2019 (Qld)*. Retrieved from https://www.legislation.qld.gov.au/view/whole/pdf/asmade/act-2019-005.

Inkster, N. (2019). The Huawei Affair and China's Technology Ambitions. *Survival, 61*(1), 105–111. https://doi.org/10.1080/00396338.2019.1568041.

Katsh, M. E., & Rabinovich-Einy, O. (2017). *Digital Justice: Technology and the Internet of Disputes*. New York: Oxford University Press.

Kim, Y. M. (2012). The Shifting Sands of Citizenship: Toward a Model of the Citizenry in Life Politics. *The Annals of the American Academy of Political and Social Science, 644*, 147–158.

Kleine, D. (2019). Development. In J. Ash, R. Kitchin, & A. Leszczynski (Eds.), *Digital Geographies* (pp. 225–237). London: Sage.

Mann, M., Devitt, K., & Daly, A. (2019). What Is (in) Good Data? In A. Daly, K. Devitt, & M. Mann (Eds.), *Good Data* (pp. 8–25). Retrieved June 10, 2019, from http://networkcultures.org/wp-content/uploads/2019/01/Good_Data.pdf.

Massey, D. (2005). *For Space*. London: Sage.

McLean, J., & Maalsen, S. (2013). A Geography of Revitalised Feminism. *Geographical Research, 51*(3), 243–256. https://doi.org/10.1111/1745-5871.12023.

Not My Debt. (2017). *Not My Debt*. Retrieved July 20, 2019, from www.notmydebt.com.au.

Oberhauser, A. M., Fluri, J. L., Whitson, R., & Mollett, S. (2017). *Feminist Spaces*. London: Taylor & Francis.

O'Shea, L. (2019). *Future Histories: What Ada Lovelace, Tom Paine, and the Paris Commune Teach Us About Digital Technology*. London: Verso Books.

Pickerill, J. (2003). *Cyberprotest: Environmental Activism Online*. Manchester: Manchester University Press.

Powell, A., & Henry, N. (2017). *Sexual Violence in a Digital Age*. Singapore: Springer.

Rawls, J. (1972). *A Theory of Justice*. Oxford: Oxford University Press.

Reddy, R. N. (2018). The Urban Under Erasure: Towards a Postcolonial Critique of Planetary Urbanization. *Environment and Planning D: Society and Space, 36*(3), 529–539. https://doi.org/10.1177/0263775817744220.

Robinson, L., Cotten, S., Ono, H., Quan-Haase, A., Mesch, G., Chen, W., et al. (2015). Digital Inequalities and Why They Matter. *Information Communication & Society, 18*(5), 569–582. https://doi.org/10.1080/1369118x.2015.1012532.

Rodino-Colocino, M. (2018). Me Too, #MeToo: Countering Cruelty with Empathy. *Communication and Critical/Cultural Studies, 15*(1), 96–100.

Ruvalcaba, O., Shulze, J., Kim, A., Berzenski, S. R., & Otten, M. P. (2018). Women's Experiences in eSports: Gendered Differences in Peer and Spectator Feedback During Competitive Video Game Play. *Journal of Sport and Social Issues, 42*(4), 295–311. https://doi.org/10.1177/0193723518773287.

Schlosberg, D. (2004). Reconceiving Environmental Justice: Global Movements and Political Theories. *Environmental Politics, 13*(3), 517–540.

Schlosberg, D. (2013). Theorising Environmental Justice: The Expanding Sphere of a Discourse. *Environmental Politics, 22*(1), 37–55.

Sohn, E. (2017). The Ice Bucket Challenge Delivers. *Nature, 550*, S113–S114. Retrieved August 1, 2018, from https://www.nature.com/articles/550S113a.

Sunstein, C. R. (2018). *#Republic: Divided Democracy in the Age of Social Media*. Princeton: Princeton University Press.

Stewart, K. (2007). *Ordinary Affects*. Durham: Duke University Press.

Tsoundarou, A. (2017, April). Technology Fail: The Centrelink Fiasco and the Impact on Vulnerable People. *Bulletin (Law Society of South Australia), 39*(3), 24–25. Retrieved July 20, 2018, from https://search-informit-com-au.simsrad.net.ocs.mq.edu.au/documentSummary;dn=800210027414593;res=IELHSS.

Van Dooren, T., Kirksey, E., & Münster, U. (2016). Multispecies Studies Cultivating Arts of Attentiveness. *Environmental Humanities, 8*(1), 1–23.

Woods, K. (2007). *Human Rights and Environmental Sustainability*. Thesis, University of Glasgow. Retrieved July 20, 2018, from http://theses.gla.ac.uk/17/1/2007woodsphd.pdf.

Yuan, L., & Wakabayashi, D. (2018, August 1). Google, Seeking a Return to China, Is Said to Be Building a Censored Search Engine. *New York Times*. Retrieved September 1, 2018, from https://www.nytimes.com/2018/08/01/technology/china-google-censored-search-engine.html.

# 5

# Decolonising Digital Technologies? Digital Geographies and Indigenous Peoples

Indigenous people are already remaking digital geographies to challenge colonial hegemonies and offer alternative narratives of, and for, Indigenous cultures, economies, societies and political processes. The question that starts the title of this chapter captures the tenuous proposition that decolonising digital technologies holds—how can digital geographies ever be decolonised given the only partial control that individuals, governments and communities have over these spaces? I am conscious of Eve Tuck's (2019, 13) reiteration that decolonisation is not a metaphor and that it is imperative to 'resist using decolonisation as a trendy term'. This chapter canvasses some ways in which Indigenous peoples are using the possibilities of digital spaces to achieve political, social, cultural and justice-based concerns. I discuss three main aspects of how Indigenous peoples are using the digital: political activism in social media and how colonial powers are challenged through creative campaigns; how identity is produced through digital engagements, and; the ways in which the digital (re)produces and amplifies aspects of colonisation that are, in turn, frequently challenged by Indigenous peoples, as well as groups from the Global South. The geographic scope of this chapter is broad, drawing on cases from settler colonial states including


J. McLean, *Changing Digital Geographies*,
https://doi.org/10.1007/978-3-030-28307-0_5

Australia and Canada, to peoples and places experiencing neocolonialism via digital technologies, including Ghana and India.

Digital spaces are facilitating new avenues for Indigenous peoples' resistance to colonial hegemonies and possibilities for community building. However, Indigenous peoples also continue to experience intense levels of racism and threats of violence as a direct result of engaging with digital spaces (Carlson and Frazer 2018a). The polar possibilities of the more-than-real are evident in how the affordances of digital technologies are regularly accompanied with high costs for Indigenous peoples. Indigenous peoples are using digital spaces to share satirical creative expression, build ad hoc publics, create timely political interventions and challenge racist stereotyping. But these are far from utopic spaces as the exposure that builds these moments and movements of resistance simultaneously also attract strong backlashes. The 'difficult work' of using digital spaces for Indigenous activism can be isolating, draining and relentless (Carlson and Frazer 2018b).

## Political Activism in Social Media by Indigenous Peoples and Organisations

> A lot of people say just that the fact that you say you're Indigenous in an open forum is a political act. And then, yeah, I did some work on the #SOSBLAKAUSTRALIA campaign, the #IndigenousDads campaign, which were all fabulous examples of the ways in which we see this global solidarity amongst Indigenous people. (Bronwyn Carlson interview, November 2018)

Aboriginal academic Professor Bronwyn Carlson points out the intensity of digital spaces in this quote and mentions two campaigns that aimed to change these frequently hostile realms through acts of solidarity. Simply declaring an Indigenous identity, demonstrating that you are surviving colonial presences, is an act of political courage and, in a country such as Australia which is only just beginning to pursue efforts at decolonisation (Saunders 2018), this resistance is impressive. But the political work does not stop at identification—proactive campaigns for

Indigenous political ends are a part of the digital fibres that make these spaces. Social media sites, for example, are never simply social sites, they are also deployed for political and cultural ends: Carlson (2013, 164) argues that 'social media is a social site but...it is also a political site where Aboriginal struggles and identities are being played out in the "new frontier."' The more-than-real is made of such multiple-use realms and this may explain the tensions they produce. On the one hand, the act of identifying as Indigenous can attract vicious personal attacks while, on the other, claiming such an identity also forges connections built through global solidarity against colonial oppression.

The more-than-real is assembled from diverse social-technical presences. One important part of the multitude of Indigenous digital political presences is @IndigenousX—a social media space that is curated by Indigenous people on a rotational basis at different times. Luke Pearson, a Gamilaroi man, founded the social media initiative in 2012 and set it up so that guest curators were encouraged to express their own identities and not subscribe to any pre-conceived structural constraints. IndigenousX is a shortened form of 'Indigenous excellence' (Sweet et al. 2013) and, as of the end of 2018, IndigenousX has 42,140 followers on Twitter and 13,302 followers on Facebook. IndigenousX also has a regular column on the Australian version of mainstream media (MSM) publication, *The Guardian*. IndigenousX is making substantial contributions to Indigenous visibility in the digital, and shifting discourses of Indigenous peoples in Australia. To give a sense of why IndigenousX has become so popular, Sweet et al. (2013) shared the reasons that people gave to justify their nomination of IndigenousX for a social media award. Supporters said that they thought IndigenousX should win 'Because it celebrates diversity whilst debunking stereotypes' and 'Because it promotes exchange among diverse Indigenous views' (Sweet et al. 2013, 106).

IndigenousX is a form of participatory journalism that enables individuals to contribute to, and benefit from, a larger audience than otherwise might be available to them without the collective. The constructive outcomes of this digitally-mediated community are, as yet, somewhat unknown and further research should examine the impact of this group. One important contribution is that IndigenousX has reframed public perceptions of Indigenous Australians and its growing presence counters histories of media exploitation of Indigenous peoples (Sweet et al. 2013). Racism towards, and stereotyping of, Indigenous peoples in Australian MSM has long been a problem and prompted calls by Langton (1993) for anti-colonial cultural criticism that aims to highlight these problems and decolonise such representations. IndigenousX could be read as an instance of self-determination and decolonisation across different social media platforms that helps to address these concerns.

Indigenous activists are using social media as a tool for various political efforts. Indigenous online collectives are growing and these target diverse political and cultural goals but are 'united by their opposition to the continuing legacy of colonialism' (Carlson and Frazer 2018b, 45). Carlson and Fraser (2018a) produced a report drawing on three years of research with Indigenous peoples about social media use and found that 79% of Indigenous people have used social media for activist purposes. Embodying a key tenet of decolonising principles, the report 'Social Media Mob: Being Indigenous online', is freely available to download. It is a wide-ranging piece of research that analyses a spectrum of uses for digital spaces. For instance, Facebook is an important space for Indigenous families to share news and keep culture strong; remotely situated communities are using mobile technologies for diverse purposes, from digital activism to maintaining ties to family and Country. Country is kin for Aboriginal and Torres Strait Islander Australians and differs to western concepts of a place of origin. Country is defined by relationships between humans and more-than-humans and is constantly co-produced by many agents, including humans and more-than-humans. As Deborah Bird Rose (1996) writes, Country is a noun and a verb, Aboriginal and Torres Strait Islanders believe that you can care for Country, worry for Country when it is sick, Country looks after you and it can be harmed by reckless acts. The more-than-real can help maintain ties to Country, expanding the way digital materialities work in diverse contexts.

Multiple scales of digital engagement drive Indigenous activism, from individuals to communities, regional areas to urban contexts, and across international borders. One example of this multi-scalar solidarity that Carlson shared in conversation was the protests against the Dakota access pipeline in the United States of America. When talking about dynamics of political activism, Bronwyn said:

> There is a collective response to it, and I could actually see that globally. So when things were happening in North Dakota, for example, Indigenous Australians were full on supporting that in the online spaces, as were people right across the world supporting us and our kind of ventures of trying to educate for change. (Interview with Bronwyn Carlson 2018)

The protests involved signing in on Facebook to Standing Rock, the site of the gathering to resist the building of the Dakota access pipeline, and attracted more than a million participants (Levin and Woolf 2016). But the success of the #NoDAPL hashtag (No Dakota Pipeline) was unclear (Whyte 2017). Kyle Powys Whyte, Potawatomi and Timnick Chair in the Humanities and A/Professor of Philosophy and Community Sustainability, Michigan State University, has written extensively on the #NoDAPL movement and expresses concern about the disconnect between enthusiasm behind a particular campaign and support for social justice for Indigenous people more broadly.

> While many allies are able to send money or show up for what they understand as "direct action," they do not know how they can advocate for indigenous peoples beyond the rare highly public issues. They do not, for example, seek to regularly pressure their political leaders to reform the U.S. government's duty to consult with tribes before construction projects. Or they are not aware of indigenous peoples facing similar struggles to that of Standing Rock who are living right next door to them. (Whyte 2017)

Whyte is critiquing the frequent gap between involvement in high-profile activism moments and support for longer-term movements that are aiming to change settler colonialism. We can see this in the Australian context with Indigenous arguments for treaty and formal national-level representation failing to gain widespread support (Davis et al. 2018). Both Whyte and Carlson have valuable insights on the efficacy of digital spaces in facilitating activism and building solidarity between First Nations, and between Indigenous and non-Indigenous peoples. It is clear that affiliations between various First Nations that pre-existed digital technologies have been strengthened by these modes of communication and that at certain times campaigns can get carriage beyond these networks. The problem with the more-than-real, in all its slippery gloss, is that deep-seated structural inequalities can remain once the 'sign in/on' to a certain campaign or petition is over, and meanwhile those who supported that moment might feel that they have done enough.

Even relatively small-scale efforts to point out flaws in stereotypes and attempts to counter racist thinking can attract backlashes that are sustained and violent. Bronwyn Carlson outlined how this happened in response to an article she wrote dispelling false understandings of the advantages that Aboriginal people receive in Australia.

> Some time ago I wrote an article saying let's talk about the free ride that everyone thinks Aboriginal people get and only a couple of days ago I got an email from someone saying, "I read your article, you're full of shit, wah, wah, what an asshole you are, blah, blah, you probably got your job because you got your degree for free, wah wah". Yeah, so it's constant, you step into the light in any way, you're going to get that. It's just… people, their true selves, come right on out.

The more-than-real is revelatory in this instance as there is no perceived need for the veneer that polite behaviour in everyday, face-to-face interactions demands, although in face-to-face interactions this may well also be absent. Bronwyn interprets digital technologies as revealing hidden inclinations, the dark sides of people that are infrequently shared, as a water diviner might reveal deep water underground. It also amplifies these 'true selves' as the means of sending messages remain available sometime after the initial interaction thanks to the more-than-real. If, for example, Bronwyn had given an unrecorded lecture on this same topic, the most likely chance for someone to engage with her argument would be in that moment. Thanks to the affordances of digital technologies, these avenues for critique remain available long after the initial sharing. The emotional intensity of the critique that Indigenous activists receive comes from the multiple ways in which these communications can happen. Anonymity can enable digital racists to make their marks repeatedly and in social media, over email, as comments on open access pieces and in personal blogs. The imagery that Bronwyn uses here is remarkable too: stepping into the light draws out aggression and recriminations, even when bringing fact-based analysis to bear on public dialogue.

## Political Activism Against Digital Colonialism: Challenging Colonial Hegemonies

Indigenous people are using digital spaces to begin to overcome structural disadvantages that are rooted in histories of colonisation and the ongoing ramifications of settler colonialism processes (Greenwood et al. 2011). Indeed, Greenwood et al. (2011, 59) argue that the opportunities to seek redress and overcome some of these impacts can be powerful: 'in terms of Indigenous development within Aotearoa New Zealand, digital technology creates less of a divide than an opportunity to repair some of the devastating landslides created by colonisation'. Māori people are seeking to reclaim the centre, rather than existing on the margins, in part through digital technologies. Greenwood et al. (2011) discuss three cases where resistance of colonisation is happening—Ngāi Tahu (tribal) language revitalisation using digital technologies, digital media used by Māori, and digital education tools. The instances that they discuss are not widespread yet but are indicative of on-the-horizon changes, they hope. For example, the 'virtual Marae', or shared digital meeting ground of Māori in Aotearoa New Zealand, may offer constructive openings for decolonisation processes.

Digital humanities scholars have been interested in decolonisation praxis for some time, recognising that settler colonialism is not an event, it is a process (Rose 2004; Wolfe 2006). Given this, the fluid qualities of digital spaces *should* enable decolonisation to occur in effective and powerful ways as processes are not immutable. Risam (2018) argues that colonisation in the digital humanities is discursive and includes the reproduction of colonial influences in the making of digital spaces, effectively centring digital ways of knowing and being on the Global North, rather than the South and with Indigenous knowledges. The materiality of digital spaces needs careful consideration too as the material changes that are required to bring about decolonisation also require attention to these aspects of settler colonial digitality.

At the same time, it is important to reflect on how decolonisation is the most recent iteration of calls for justice by and for Indigenous peoples. Tuck (2019) argues that decolonisation is not an end point, but the latest way of framing ongoing processes which aim to reshape relationships between Indigenous peoples and settler institutions. Given this long history and the 'nowness' of decolonisation, it is worth considering Aboriginal academic Langton's (1993) analysis of representation of Indigenous peoples in MSM. Aboriginal academic Marcia Langton (1993, 8) asked 'Can we ever decolonise Australian institutions? Can we decolonise our minds? Probably not. But we can find ways to undermine the colonial hegemony'. Indigenous activists may not be seeking decolonisation of digital spaces at all times given these structural impediments and the sheer messiness of the more-than-real.

When Bronwyn and I talked about decolonisation of digital spaces, the fault lines in capitalist and colonial structures appeared in stark relief:

*Bronwyn*: Decolonization is an interesting thing. So, we're talking about a capitalist corporate model that we're functioning in and wondering how we can decolonize that. So that, in itself, is a bit of an issue.

*Jess*: Yeah.

*Bronwyn*: But indigenous people say, "Hell no, we're not getting off social media, 'cause this is our place to participate in the world." So, is it that to really decolonize those spaces, they would have to be taken out of the corporate capitalist scene and then do we give them to governments to run?

That's a terrifying thing, seeing as they [governments] are the storm troopers of capitalism. You know, I could imagine Dutton[1] on there and

[1]Peter Dutton and Tony Abbott, politicians and members of the hard-right faction of the Federal Liberal Party, have shared arch conservative views on Indigenous peoples while in office.

> having a say and Abbott thanking us all for the invasion, putting up with it. So, it is a worry. The very space itself, I'm not sure that can be decolonized, without the good will of those who benefit from this space, but certainly people can make some attempts to think a little differently online. And to think in a more just way and don't lose your humanity. What are the things that are important here? What I say might impact somebody's world, but I don't think we're there yet. And so I don't think we're there yet… (Interview with Bronwyn Carlson, October 2018)

The similarities between Carlson's and Langton's arguments about undermining the colonial hegemony and doing things in a 'more just way' are important here. While Langton was discussing MSM and Carlson digital spaces, both recognise the limits and possibilities that technologies afford in terms of changing broader social, cultural and political contexts for Indigenous people. Governments are not easily trusted given their roles in settler colonial societies and corporations may not have the capacity, profit motive or interest to play active roles in decolonisation.

## Digital Storytelling as Identity Building and Healing

Indigenous peoples have been using communication technologies to share a broad range of narratives, including complicating simple stories of marginalisation with humour and satire. For instance, Tracey Moffatt's film *Nice Coloured Girls* (1987) centres the agency of young Aboriginal women and overturns stereotypes of oppression in an experimental way. Images of colonisation, including the First Fleet landing, are overlaid with young Aboriginal women exploiting their 'Captain', an older white man, and eventually taking his money and sending him home (Jennings and Hollinsworth 1987; Sullivan 2018). Sullivan (2018) convincingly argues that Moffatt's film repositions Indigenous sex workers as agential as they exploited an opportunity, when there were very few, to gain economic advantage.

In digital spaces, similar tactics have been used to subvert paradigms of negative/positive Aboriginality (Langton 1993). For instance, Berglund (2017) describes the work of the 1491s, a group that uses satire to challenge colonial thinking and practices. The 1491s do not deny the gravity of the consequences of colonisation but they do want people to acknowledge their agency and to recognise the variety of ways in which Indigenous peoples are surviving and thriving, despite difficult structural forces. On Twitter, 1491s describe themselves in the following way:

> The 1491s is a sketch comedy group based in the wooded ghettos of Minnesota and buffalo grass of Oklahoma. They are a gaggle of Indians chock full of cynicism.

Over 20,000 people follow their Twitter account and many of their videos on YouTube have over a million views. For instance, 'Singing Lessons' with 1.1 million views shows a Native American elder instructing a younger Indian man how to sing in a 'traditional' way. The younger man is nearly brought to tears by the Elder's critique and is finally deemed to have sung properly, 'with passion'. Berglund (2017) understands the humour shown in the 1491s work as a form of work that is seeking transformation. Humour is a political strategy here in that it 'recognises that humour is a tactic used to move people, to change attitudes, to critique behaviours, and to lead to change' (Berglund 2017, 5). The self-reflexivity of what it means to do cultural performances properly belies the negative/positive duality so often put forth in Indigenous representations and offers a way to think differently about settler colonial relations.

Sharing stories of multiple sorts in digital spaces can be empowering and may work to shift stereotyping binaries of positive/negative Indigeneity. Such digital storytelling may even have decolonising effects in terms of helping people overcome damaging social, political and cultural effects of settler colonialism. For instance, Adelson and Olding (2013) describe digital storytelling as a potentially healing process as Indigenous people are able to rewrite imposed colonial narratives and give voice to their own lived experiences. A good example of this is the 'What's ya Story' app that was developed by Indigenous young people, researchers from the University of Melbourne, and guided by the Korin Gamadji Institute (KGI) at the Richmond Football Club, Victoria. Edmonds et al. (2014) describe the project as involving 'Indigitalising' technologies, customising apps and other digital technologies to support Indigenous identities through facilitating yarns and other Indigenous ways of knowing and doing. One way that the app achieves this is the use of 'deadly' in place of the common place 'like' to signal support for a post. The stories that are shared through the app are grounded through connections to Country and these are explicitly brought into the architecture of the app. Further, participant stories could be created and shared across a range of media, including songs, poetry, text, using maps and more. The app prompted reflections from participants about the possibility of cross-generational connections and novel ways to record important narratives. Alex, a participant in the research, offered the following powerful reflections on the app:

> The kids have a responsibility to help their Elders be able to put their stories on this app they could actually sit down with them; instead of their Elder typing it in on the phone, [kids] can type it in on the phone for them. And, having that digital literacy and supporting an Elder is probably a great benefit. [Kids] could actually sit down there and do it for [Elders] but [also] just listen to what their [Elder's] story is and put it down for them. So that's sharing the burden of being able to pass on that story and using two different generations to produce something that can go further. (Edmonds et al. 2014, 37)

Building capacity across generations and 'sharing the burden' of communicating stories are two important aspects of the What's ya Story app according to Alex and these aspects could be read as healing moments for younger Indigenous people and their Elders.

Creative use of social media can also facilitate Indigenous resistance to racism and stereotyping while simultaneously building Indigenous identities. Ad hoc publics form among people and institutions who may not follow each other and yet can build momentum by using hashtags to create spontaneous communities at key political moments

(Bruns and Burgess 2011). Ryan Griffen, creator of the series *Cleverman*,[2] an excellent superhero-driven series weaving Aboriginal stories with futuristic worlds, has been using Twitter to start conversations that counter misrepresentations of Indigeneity. One of these efforts was the #IndigenousDads Twitter meme that invited Indigenous Australians to share narratives and images of their fathers to counter the racist trope offered in a Bill Leak penned cartoon (Ford 2017). The conservative newspaper *The Australian* published a cartoon depicting a drunk Aboriginal man not knowing the name of his son when confronted by a police officer. Ryan Griffen started the ad hoc public by writing 'Not only do I know my sons name but I named a superhero after him. #IndigenousDads #Cleverman'. His tweet started a conversation that allowed diverse narratives of Indigenous fatherhood to appear on social media and counter the negative characterisation offered by *The Australian*. Davey (2016) reports on the many shared stories under #IndigenousDads and some of these are collated in Table 5.1. These tweets were accompanied by evocative photos of strong, active Indigenous men—directly challenging the tired image of a wasted Indigenous man that Leak offered.

The #Indigenousdads ad hoc public facilitated a strong response to a racist MSM message and these powerful and spontaneous interventions are a recurring trope in the more-than-real. These counter-discourses undermined the stereotyping at play in MSM and asked people to reflect on the assumptions embedded in the cartoon, as well as broader stereotypes about Indigenous parenting in a humorous and assertive way. Indigenous fathers used social media to rewrite the imposed narrative and in doing so asserted their power and undermined the colonial reach.

[2] *Cleverman* aired on Sundance television on 1 June 2016 and the next day on Australian Broadcasting Corporation television.

**Table 5.1** A sample of the #Indigenousdads offerings

| Twitter handle | Occupation of Tweeter | Tweet | Likes | Retweets |
|---|---|---|---|---|
| @chrissarra | Rugby League Commissioner | 'That's me & my youngest.. And that's a FISH!!!!! Not a beer can!!! You CANNOT stereotype me!!! #IndigenousDads | 698 | 252 |
| @amycarolcollins | Unknown | My partner who is a wonderful father that does everything with his children #IndigenousDads | 71 | 45 |
| @kristal_kinsela | Aboriginal Business Woman of the Year 2017 | 2 time Olympian, Vietnam Vet, Father and all round hero my uncle John Kinsela #indigenousdads | 279 | 153 |
| @jpjanke | Co-chair of National NAIDOC Week Committee | Don't even drink beer and know all the names of my sons. That 'cartoon' labelled me otherwise. #IndigenousDads | 685 | 1410 |
| @azzanagas7 | Unknown | Was proud to hear #Indigenousdads trended last night. here's my contribution of my beautiful jarjums & I #love | 368 | 147 |

## Neocolonial Digital Technologies

There is a deepening and broadening of digital engagement in everyday lives—as well as possibilities for challenging structural dynamics that reproduce inequalities and intensify corporatisation of a range of digital geographies. We know that uneven geographies of representation and participation in the digital are commonplace (Graham et al. 2015) as the Global North dominates digital information technologies, both in terms of making and being the subject of most digital data. Further, in Google searches, the Global South is more frequently represented *by* the North rather than *self-represented*: this form of digital hegemony skews digital data and could potentially misrepresent the Global South (Ballatore et al. 2017). The issues with corporate expansion from the Global North via digital geographies are complex and loaded with capitalist inequities, as demonstrated by Facebook broadening its presence in the Global South. Facebook's 'Free Basics' app was developed for people to use mobiles to access some websites and services that aren't data intensive. It has been widely critiqued, including by Ellery Biddle, advocacy director of Global Voices:

> Facebook is not introducing people to open internet where you can learn, create and build things. It's building this little web that turns the user into a mostly passive consumer of mostly western corporate content. That's digital colonialism. (quoted by Solon 2017)

The low levels of data required to access these sites is a function of the lack of images and videos therein; Free Basics can be browsed without charge and includes Facebook and country-specific information. Qualitative research on the Free Basics app by Gebhardt (2016) involved research with the app users in Ghana and concluded that users engage with these services in divergent ways to those that Facebook expects. Gebhardt (2016) found that users were conscious of the fact that Free Basics was a conduit for expansion of telecommunications and digital companies. Research participants in Ghana were irritated at the data exclusions that they encountered while using these digital technologies and would simply end up not using Free Basics. Rather than being controlled through a sort of digital colonialism, Ghanaians abandoned the app. The digital colonialism that Ellery Biddle observes in Free Basics is subverted by users who do not want to engage with these spaces.

In South Africa, Romanosky and Chetty (2018) find that a different picture emerges with Free Basics users. People have been using the digital spaces provided by Free Basics for longer periods of time than they otherwise would due to the 'free' framing of those sites, and people are not entirely sure of what 'free' means in South Africa according to Romanosky and Chetty (2018); the limited range of digital software that users are able to use and the avalanches of advertisements are the costs of these free services. Digital colonialism in this context has deep impacts as an absence of net neutrality is obvious and the contingent expansion of digital spaces is delivering poor outcomes for people.

Neocolonial digital technologies are a part of transformations of cities around the world also thanks to smart cities interventions. Smart cities technologies can range from digital urban forests (Phillips and Atchison 2018) to garbage bins communicating their capacity (Medvedev et al. 2015). Smart cities are generally associated with techno-optimism but, as Luque-Alaya (2019) notes, the unevenness of economic prosperity within current cities will be deepened with the overlayering of digital platforms. While there are many unknowns about the implications of big data applications in smart cities contexts, from megacities to global cities, it is fairly certain that introducing digital transformations will exacerbate multi-scaled inequalities. For example, Datta (2019) analyses the framing of smart cities in India and dissects the enumerations, articulations and breaches of these material and discursive processes. The enumerations of smart cities refers to how states coerce citizens into the smart cities fold, articulations of smart cities include the remaking of inequities from analogue contexts in the digital and breaches appear when alternative and possible cities are glimpsed as people rupture the smooth surfaces of smart cities. The decolonised urban age, in whatever form that emerges, will involve citizens engaging with material and digital worlds and demanding further critical geographic scholarship (Datta 2019), building on Strengers (2013) critiques of utopic thinking that smart cities rhetoric tends to offer.

## Conclusion

Grappling with the complexities of digital and non-digital spaces can involve contesting narratives of Indigenous ways of being. For people like Jason Edward Lewis who started Aboriginal Territories in Cyberspace, a digital technology company that makes VR, this is an ongoing problem 'We are, sadly, at the stage where simply stating Indigenous presence is a still a radical thing, in both the real world and within virtual spaces' (quoted in Keating 2018, paragraph 15). According to Lewis, any presence at all is an improvement on the paucity of Indigenous representations currently available.

Digital geographies offer avenues for change as well as sources for entrenching injustices; it is how people use and remake particular technologies that render these devices and spaces as oppressive or liberatory. Just as the category 'Indigenous' is multiple and situated, so too are the ways in which Indigenous peoples co-produce digital geographies. An important strength of digital spaces for some—the affordance of anonymity—is a challenge for many Indigenous people as building and maintaining Indigenous communities is easier with visibility, identification and connection between Indigenous peoples. For Bronwyn Carlson, this tension has been a key theme of her research:

> To be Indigenous, you know, you have to openly identify. And the community recognizes that identification, and in some ways that there's a formative element to it, because you have to be seen to be Indigenous. So it's this kind of package and being anonymous online doesn't solve that. The fact is you want people to know who you are and that you're Indigenous. So yeah, how do they do that online, that was interesting to me. There are all these kind of things that are a little bit different to how non-Indigenous people might use the space that I found fascinating.

Unlike some groups who may find safety and possibility in digital anonymity, for example gay and lesbian teenagers (Craig and McInroy 2014), some Indigenous people face a double dilemma of having to share identities for recognition and community acceptance but at that moment also risk exposing themselves to vitriol and aggression.

Decolonising digital technologies is not a straightforward proposition, and neither is it necessarily something that is desirable or feasible. For corporations to 'decolonise', some form of government intervention may be required and this would be anathema for many Indigenous peoples. The problem of socially and culturally damaging behaviours in digital technologies is not unique to Indigenous peoples but due to the greater use of tools such as social media Indigenous peoples may be experiencing higher levels of abuse. Finding ways to undermine the colonial hegemony is an ongoing focus for Indigenous peoples, communities, and insitutions who are using digital technologies. While MSM may continue to be a source of stereotyping and racism towards Indigenous peoples, social media is providing a platform for Indigenous people to self-determine and challenge such narrow views. At the same time, social media is no panacea as it also affords racist and derogatory behaviours, and there are sometimes high personal costs associated with engaging with the digital. The risks associated with putting oneself in the public sphere are great for Indigenous peoples, but the benefits are also noteworthy.

The skewed representation of the Global South by the Global North in digital spaces reproduces old dilemmas that can be, and have been, countered by efforts to critique neocolonial practices. Digital corporations have a responsibility to ethically engage with people, places and communities, even while having economic agendas as their main priority; people, communities and institutions are pushing back against what they perceive as unjust digital impositions. The more-than-real cuts through multiple scales when considering colonial processes, Indigenous realities and global forces of change. The contradictions inherent to digital geographies are an ongoing focus for Indigenous artists, academics, activists and communities as they work to redefine digital geographies.

## References

Adelson, N., & Olding, M. 2013. Narrating Aboriginality On-line: Digital Storytelling, Identity and Healing. *The Journal of Community Informatics, 9*(2). Retrieved from http://ci-journal.net/index.php/ciej/article/view/740/1004.

Ballatore, A., Graham, M., & Sen, S. (2017). Digital Hegemonies: The Localness of Search Engine Results. *Annals of the American Association of Geographers, 107*(5), 1194–1215. https://doi.org/10.1080/24694452.2017.1308240.

Berglund, J. (2017). 'Go Cry Over Someone Else's Tragedy': The YouTube Activism of the 1491s. *Australasian Journal of Information Systems, 21*. http://dx.doi.org/10.3127/ajis.v21i0.1581.

Bruns, A., & Burgess, J. E. (2011). *Proceedings of the 6th European Consortium for Political Research (ECPR) General Conference 2011, University of Iceland, Reykjavik*. Retrieved July 20, 2018, from https://eprints.qut.edu.au/46515/.

Carlson, B. (2013). The 'New Frontier': Emergent Indigenous Identities and Social Media. In M. Harris, M. Nakata, & B. Carlson (Eds.), *The Politics of Identity: Emerging Indigeneity* (pp. 147–168). Sydney: University of Technology Sydney E-Press.

Carlson, B., & Frazer, R. (2018a). *Social Media Mob: Being Indigenous Online*. Sydney: Macquarie University. Retrieved June 10, 2019, from https://research-management.mq.edu.au/ws/portalfiles/portal/85013179/MQU_SocialMediaMob_report_Carlson_Frazer.pdf.

Carlson, B., & Frazer, R. (2018b). Yarning Circles and Social Media Activism. *Media International Australia, 169*(1), 43–53. https://doi.org/10.1177/1329878X18803762.

Craig, S., & McInroy, L. (2014). You Can Form a Part of Yourself Online: The Influence of New Media on Identity Development and Coming Out for LGBTQ Youth. *Journal of Gay & Lesbian Mental Health, 18*(1), 95–109. https://doi.org/10.1080/19359705.2013.777007.

Datta, A. (2019). Postcolonial Urban Futures: Imagining and Governing India's Smart Urban Age. *Environment and Planning D: Society and Space, 37*(3), 393–410. https://doi.org/10.1177/0263775818800721.

Davey, M. (2016). Australians #IndigenousDads and Their Children Take to Twitter Over Bill Leak Cartoon. *The Guardian*. Retrieved December 10, 2018, from https://www.theguardian.com/australia-news/2016/aug/07/indigenousdads-and-their-children-take-to-twitter-over-bill-leak-cartoon.

Davis, M., Dixon, R., Appleby, G., & Pearson, N. (2018, Autumn). The Uluru Statement. *Bar News: The Journal of the NSW Bar Association* 41. Retrieved May 5, 2019, from http://classic.austlii.edu.au/au/journals/NSWBarAssocNews/2018/20.pdf.

Edmonds, F., Rachinger, C., Singh, G., Chenhall, R., Arnold, M., de Souza, P., & Lowish, S. (2014). *What's Ya Story: The Making of a Digital Storytelling Mobile App with Aboriginal Young People*. Sydney: Australian Communications Consumer Action Network. Retrieved May 10, 2019, from https://minerva-access.unimelb.edu.au/handle/11343/42894.

Ford, F. (2017, June). Screen Dreaming in 'Cleverman': Reimagining Indigenous Identities. *Screen Education, 85,* 26–35. Retrieved December 10, 2018, from https://search.informit.com.au/documentSummary;dn=987899946898441;res=IELLCC.

Gebhart, G. (2016), Zero-Rating in Emerging Mobile Markets: Free Basics and Wikipedia Zero in Ghana. In *Proceedings of the Eighth International Conference on Information and Communication Technologies and Development* (p. 16). ACM. Retrieved May 14, 2019, from https://digital.lib.washington.edu/researchworks/bitstream/handle/1773/37299/Gebhart%20ICTD%20author%20final.pdf?sequence=1.

Graham, M., De Sabbata, S., & Zook, M. A. (2015). Towards a Study of Information Geographies: (Im)mutable Augmentations and a Mapping of the Geographies of Information. *Geo: Geography and Environment, 2*(1), 88–105. https://doi.org/10.1002/geo2.8.

Greenwood, J., Te Aika, L. H., & Davis, N. (2011). Creating Virtual Marae: An Examination of How Digital Technologies Have Been Adopted by Māori in Aotearoa New Zealand. In P. Randolph Leigh (Ed.), *International Exploration of Technology Equity and the Digital Divide: Critical, Historical and Social Perspectives* (pp. 58–79). Hershey, PA: IGI Global. https://doi.org/10.4018/978-1-61520-793-0.ch004.

Jennings, K., & Hollinsworth, D. (1987). Ways of Seeing and Speaking About Aboriginal Women: ii. Shy Maids and Wanton Strumpets. *Hecate, 13*(2), 129–133.

Keating, C. (2018). *Indigenous People Are Decolonizing Digital Worlds*. The Medium. Retrieved May 8, 2019, from https://medium.com/s/story/indigenous-peoples-are-decolonizing-virtual-worlds-fbe2d8f933de.

Langton, M. (1993). *Well, I Saw It on the Television and I Heard It on the Radio*. Sydney: Australian Film Commission.

Levin, S., & Woolf, N. (2016). A Million People 'Check in' at Standing Rock on Facebook to Support Dakota Pipeline Protesters. *The Guardian*. Retrieved December 7, 2018, from https://www.theguardian.com/us-news/2016/oct/31/north-dakota-access-pipeline-protest-mass-facebook-check-in.

Luque-Ayala, A. (2019). Urban. In J. Ash, R. Kitchin, & A. Leszczynski (Eds.), *Digital Geographies* (pp. 24–35). London: Sage.

Medvedev, A., Fedchenkov, P., Zaslavsky, A., Anagnostopoulos, T., & Khoruzhnikov, S. (2015). Waste Management as an IoT-Enabled Service in Smart Cities. In *Internet of Things, Smart Spaces, and Next Generation Networks and Systems* (pp. 104–115). Cham: Springer.

Phillips, C., & Atchison, J. (2018). Seeing the Trees for the (Urban) Forest: More-Than-Human Geographies and Urban Greening. *Australian Geographer*. https://doi.org/10.1080/00049182.2018.1505285.

Risam, R. 2018. Decolonizing the Digital Humanities in Theory and Practice. In J. Sayers (Ed.), *The Routledge Companion to Media Studies and Digital Humanities* (pp. 78–86). New York: Routledge. Retrieved June 9, 2019, from https://digitalcommons.salemstate.edu/cgi/viewcontent.cgi?article=1006&context=english_facpub.

Romanosky, J., & Chetty, M. (2018). *Understanding the Use and Impact of the Zero-Rated Free Basics Platform in South Africa*. In *Proceedings of the 2018 CHI Conference on Human Factors in Computing Systems (CHI '18)* (Paper 192, 13 pages). New York, NY: ACM. https://doi.org/10.1145/3173574.3173766.

Rose, D. B. (1996). *Nourishing Terrains: Australian Aboriginal Views of Landscape and Wilderness*. Canberra: Australian Heritage Commission.

Rose, D. B. (2004). *Reports from a Wild Country: Ethics for Decolonisation*. Sydney: UNSW Press.

Saunders, I. (2018). Post-colonial Australia: Fact or Fabrication? *New: Emerging Scholars in Australian Indigenous Studies, 2–3*(1), 56–61. http://dx.doi.org/10.5130/nesais.v2i1.1474.

Solon, O. (2017). 'It's Digital Colonialism': How Facebook's Free Internet Service Has Failed Its Users. *The Guardian*. Retrieved August 10, from https://www.theguardian.com/technology/2017/jul/27/facebook-free-basics-developing-markets.

Strengers, Y. (2013). *Smart Energy Technologies in Everyday Life: Smart Utopia?* London: Palgrave Macmillan.

Sullivan, C. (2018). Indigenous Australian Women's Colonial Sexual Intimacies: Positioning Indigenous Women's Agency. *Culture, Health & Sexuality, 20*(4), 397–410. https://doi.org/10.1080/13691058.2017.1349930.

Sweet, M., Pearson, L., & Dudgeon, P. (2013). @IndigenousX: A Case Study of Community-led Innovation in Digital Media. *Media International Australia, 149*(1), 104–111. https://doi.org/10.1177/1329878X1314900112.

Tuck, E. (2019). Losing Patience for the Task of Convincing Settlers to Pay Attention to Indigenous Ideas. In Introduction by L. T. Smith, E. Tuck, & K. W. Yang (Eds.). (2018). *Indigenous and Decolonizing Studies in Education: Mapping the Long View* (pp. 13–16). New York: Routledge.

Whyte, K. P. (2017, January 6). *Five Reasons Why the North Dakota Pipeline Fight Will Continue*. The Conversation. Retrieved August 10, 2018, from https://theconversation.com/five-reasons-why-the-north-dakota-pipeline-fight-will-continue-in-2017-70782.

Wolfe, P. (2006). Settler Colonialism and the Elimination of the Indigenous. *Journal of Genocide Research, 8*(4), 387–409.

# 6

# Changing Climates Digitally: More-Than-Real Environments

The consumption of digital technologies, especially in the Global North, is fundamentally environmentally unsustainable (Bates et al. 2015; Berkhout and Hertin 2004; Kleine 2019). At the same time, digital activism is transforming possibilities of environmental sustainability, including facilitating more efficient and effective digital activism *and* a focus on how to make digital spaces more sustainable. This chapter focuses on how climate action in Australia relies on digital interventions in multiple guises: from the crowdfunding that created the Climate Council to the digital tools that enable the visibility of other environmental action groups.

As a critical geographer, I take the stance that space is material and discursive, and this chapter focuses on what may be seen as materially produced and discursively driven in the context of more-than-real environments. The emotional landscapes that drive movement building are well known and widely reported—mostly around how outrage propels action (Asenbaum 2018; Spring et al. 2018). This chapter examines how hope, anger and grief are also key emotions that contribute to changes in digital spaces and the ways in which environmental NGOs attempt to channel and harness these emotional trajectories. Drawing


J. McLean, *Changing Digital Geographies*,
https://doi.org/10.1007/978-3-030-28307-0_6

on Asenbaum's framing of cyborg activism, this chapter considers how pragmatic use of digital spaces allows environmental activists to persist, relying on deep entanglements with sometimes problematic digital technologies. The structural conditions that impede or allow such action are brought to light in conversations about strategy, as those charged with activating people for environmental action constantly have to negotiate top-down shifts in digital geographies.

This chapter begins by telling a story of energy contestation in social media involving Scott Morrison (the Prime Minister of Australia as of 2019), Mike Cannon-Brookes (a tech billionaire), Australian climate activist communities and other Australians. The Prime Minister wraps up themes of patriotism, authenticity and rationality in emotional language as he appeals to people's interest in 'fair dinkum power' while Cannon-Brookes responds with satire to invert the implied insult to renewables as inauthentic. Joining the multiple efforts of other environmental NGOs seeking climate change action, Cannon-Brookes has now begun a movement for achieving much higher levels of renewable energy in Australia. I then explore how climate action is proceeding in Australia from the perspectives of those already working within that sector. The way emotion is mobilised to activate change is demonstrated in this analysis, as well as critical reflection about the strengths and weaknesses of digital campaigning. A hybrid form of public is evident in this analysis, as institutions emerge through blends of ad hoc publics (Bruns and Burgess 2011) and planned movements. Digital activism is, after all, inherently hybrid. Last, I analyse how people who are seeking transformations in, and of, digital spaces frame more-than-real environments.

Climate change action in Australia is the focus of this chapter due to the dominance of this environmental issue in terms of activism and the current impacts of climate change, as well as the high risks of future warming. Facilitating the uptake of renewable energy is a strong component of this action. As of 2018, 2 million solar power photovoltaic units have been installed on homes around Australia (Hannam 2018) and non-government organisations are advocating solar installation as a way to contribute to renewable energy growth and carbon emissions reductions. Despite a lack of strong federal government commitment to climate change action, communities are taking material action that makes a difference (Klein 2018; McLean 2016). This chapter examines how these diffuse activist efforts are contributing to substantial shifts in the diverse economies that people and institutions build. The paradoxical qualities of the spatial distribution of solar partly stem from the high uptake of solar power by lower income householders. Households with less disposable income are placing high value on sustainability and committing to such in large numbers. Micro-geographies of disruption in the Anthropocene reveal important and surprising links between desire—to do something—and economic rationalist discourse and practice.

## Playing with 'Fair Dinkum' Power

In October 2018, the newly minted Prime Minister of Australia, Scott Morrison, shared a Facebook video about the way his government was going to change energy prices for their constituents. The one minute video starts with the declaration that 'We've got to get electricity prices down'. The Prime Minister uses an anecdote of an Australian couple, one a returned services officer, and states that their electricity bill is too high. The plan is to get 'electricity companies in line' with new legislative measures and assures that they won't be able to 'rip you off simply by you being a loyal customer'. The next step will be new penalties to stop 'price gouging' and last:

> We're going to force them to put more fair dinkum, reliable energy power into the system. Renewables are great. But we are also needing the reliable power when the sun isn't shining and the wind isn't blowing. That's what keeps the lights on. (Morrison 2018 on Facebook)

Before Morrison was Prime Minister, he famously brought coal into Federal Parliament as a prop (Hamilton 2017) and made a speech about the dangers of 'coalphobia', mocking those who want to close coal power stations, and arguing that there is no reason to fear the 'black rock'. So while in this video about electricity prices the Prime Minister does not explicitly advocate coal power, it is definitely in the background as a form of 'fair dinkum power'. And in the foreground is an undermining of renewables—explicitly, solar and wind power—as unreliable and needing more authentic, more 'reliable' support. The term 'fair dinkum' is a colloquial phrase that is defined by the Macquarie Dictionary as meaning earnest, true, genuine and 'typically Australian' (Macquarie Dictionary 2018). The assignation of some forms of power generation as typically Australian, and others as not, brings curious debates around nationhood and resource management together. The Liberal party, which Morrison leads, is purported to value economically rational policies and practices but this coupling of patriotism with fossil fuel use for energy production suggests an ideological position that is far from rational. Similar to other conservative leaders in wealthy nations such as Trump (Brabazon et al. 2018), Morrison conjoins nationalistic fervour with reticence on climate change action, joining the likes of political leaders in Britain, Denmark and Italy (Lockwood 2018).

The Facebook video of Morrison exhorting fair dinkum power received over 1200 comments, about 1000 likes, and 287 shares; the average post on the Prime Minister's page attracts 50–100 comments and 30–100 likes. The video gained the attention of Mike Cannon-Brookes, co-CEO of Atlassian, a software company, and he retweeted it with a critique and promise to help Australia become a renewables exporter (Fig. 6.1).

Anger (the red-faced anguished emoji) and inspiration ('we need a rallying cry') are motivators for Cannon-Brookes here as he adopts a common activist strategy of inverting the PM's original meaning of 'fair dinkum power' in a subversive move, similar to other protest strategies (Penney 2015). By repositioning the framing of renewable energy as the fair dinkum power Morrison desires, Cannon-Brookes is satirising the way the PM mobilises colloquialisms in energy discourse. Rather than arguing against the patriotic positioning that Morrison put forward for fossil fuels, he turned the language and intention around and has begun a 'fair dinkum power' campaign for renewables. The campaign involves an online petition to gather support for affordable renewables and a manifesto that attests: 'We believe our power can be clean, cheap, reliable and Australian' (Fair Dinkum Power 2018). Merchandise including t-shirts and bumper stickers for this pro-renewable energy campaign were available within four days of the website and campaign launch.

**Fig. 6.1** Mike Cannon-Brookes' response to PM video on energy companies and production

The sociopolitical dynamics surrounding the 'Fair Dinkum Power' Cannon-Brookes campaign and the PM is not entirely new: Elon Musk famously intervened in South Australia's energy crises by providing the technological know-how and infrastructure to build a battery for storage of renewable energy (Hunt 2017) in a social media exchange with Mike Cannon-Brookes. The South Australian government took Musk up on that offer and has paid for the battery as it was installed ahead of schedule. The more-than-real is formed by discursive and material interventions of cyborg activism. Political negotiations with respect to climate change action are not limited to the formal halls of politics but are taking place in digital geographies and beyond.

## Power, Emotions and Environmental NGOs in the More-Than-Real

Digital action in environmental NGOs involves a mixture of opportunistic strategy and long-term planning. Working with emotion and affect is a key part of the strategies that these activists employ. From an interview with a Community Engagement officer of one of the largest climate change action NGOs in Australia, it was clear that emotions play a central role in their campaigning:

*Jess*: Does consideration of emotions play a part in how you run online campaigns?

*Clara*: Definitely. For us, most of our reactive campaigns (they're generally the campaigns that, for us, perform the best, particularly in terms of fundraising) are based on moments. As a general rule, we find these moments, and we can tap into people's outrage in particular. That would be, for us, the emotion that performs the best. So, whether that's the government holding back data on Australia's rising emissions, cutting funding to CSIRO, or climate censorship on the state of the Reef with the UNESCO report. Those are things we've had as fundraising moments, and they're moments when the community is engaged and is passionate and is most ready to take action. **We definitely try and ride that wave and, I suppose, encourage those emotions to be played out.** Of course, you can't always plan for those moments, and so we also try and generate emotions, and the situations still create a similar effect. (Interview with Environmental NGO founder, November 2018, emphasis added)

This two-way approach, either drawing on already circulating outrage or anger while also looking for ways to generate affectual responses among supporter groups, clearly shows flexible and creative modes of working in the more-than-real. Jasper (1998) describes the mixture of ways in which emotions interact with protest and argues that if anger is a part of activism it may not necessarily lead to irrational acts. The long-term affiliations or aversions to particular groups is a key idea that Jasper (1998) forwards for how emotions underpin political action. People traditionally have been committed to causes and/or activist groups for sustained periods of time. Political action is propelled by these deeper, more sustained affects, according to Jasper, but now it seems that the rise of issue publics in political life within digital spaces challenges this duality. Emotional responses are just as likely to propel social and political changes as shown in the creation of the Climate Council. The Council's emergence came from anger and disappointment that culminated in a transformed institution. Issue publics on social media tend to emerge from individuals contributing their personal narratives to campaigns that create solidarity (Kim 2012).

The mobilisation of people through anger is not the end of the process for activists working with digital technologies. Rather, it is often a starting point that then merges into hope and, finally, solutions. So generating anger or outrage, or tapping into surges of anger or outrage, is never an end point in and of itself but a step along a trajectory of actions. As an Environmental NGO founder attests, this is not a new approach either. Rather, the anger-hope-action frame is a tried-and-true method from pre-digital activism:

> *Jess*: The relation between anger, hope, and action is an organizing principle of your activism then?
>
> *Maz*: Yeah. That's a very old standard organizing framework or framework for communication to volunteer, or somebody to take action but also to 'like' a campaign, there's other ones as well but that's just an example. A very simple and widely used and understood one. (Environmental NGO founder interview, October 2018)

Power is produced, negotiated and contested in fluid ways in different forms of digital action. The capacity to harness attention and maintain authority within issue publics in digital spaces does not always require individuals or groups to have powerful roles in other contexts. As Sandover et al. (2018) note in their analysis of social media and the badger cull in England, an instance of environmental activism in digital spaces, individuals gained power through original contributions to the outcry. So people who may not hold politically powerful roles elsewhere were able to garner significant attention in digital spaces due to their novel digital engagement.

Scientists such as Professor Lesley Hughes, a biologist at Macquarie University who is also a Councillor for the Climate Council and a Director of the World Wildlife Fund for Nature (WWF), has seen first-hand how emotions and digital engagement are connected. As one of the founding Councillors of the Climate Council who was involved in the transformation of that institution from a government funded to independent entity, Hughes was closely involved in a moment of simultaneous anger and hope in Australian environmental action:

*Jess*: I'm just wondering, for a lot of people there's outrage that was a part of the process of saying "Yes we want to support the creation of the Climate Council and we'll keep supporting you". For a lot of people who don't engage in climate action that seems to have been a motivation to engage and counter feelings of apathy ?

*Lesley*: Climate change communicators walk a fine line. On the one hand as a scientist I feel compelled to explain the science as we currently understand it – and that gives a fairly bleak picture of our potential future. On the other hand, we know that talking about catastrophe all of the time tends to cause people to shut down - it's just too relentlessly depressing to take it all in. So finding that balance to get people concerned but also remain hopeful and active is the challenge. (Lesley Hughes interview, November 2018)

The tightrope walk, between catastrophic thinking and hope that interventions may be worthwhile, is a complicated one to take as an individual, let alone when attempting to encourage individuals and communities to walk along with you. The similar relationship between anger and hope is mentioned in Head's (2016) treatise on grief and hope in the Anthropocene. Head (2016) draws on Sara Ahmed's forwarding of emotional engagement who argues that 'we must persist in explaining why our anger is reasonable, even in the face of others who use this anger as evidence of poor reason' (Ahmed 2004, p. 177 in Head 2016, 34). Anger may be reasonable but it also must, in the minds of environmental activists and scientists, translate into meaningful action. And there lies a challenge of digital spaces: anger flares up, can be and is often contagious, generates digital commentary and then can either produce substantive interventions—or nothing.

The work of explaining why anger is reasonable can be debilitating and digital action seems to provide an avenue for compression of this shift, from anger to action via hope. The affordances that digital geographies provide include pathways for effective engagement. The more-than-real provides a straightforward process for channeling anger and mobilising groups in multiple ways. From building a financial base for effective action, to supporting local-scale and individual-scale interventions, digital tools provide avenues for moving from anger to hope. All environmental NGO leaders that I spoke with indicated how powerful anger is within digital spaces, and beyond, in terms of: building communities; sustaining financial support, and; making effective issue publics. For example, Maz, an Environmental NGO founder, described the reactive politics that form in response to a climate change sceptic leader such as Tony Abbott:

*Jess*: I'm curious as to whether in your environmental NGO, or other organisations that you've worked in, considerations of emotions play a role in how you run campaigns?

*Maz*: Yes. Absolutely. We know that emotions play a critical role in moving people to act and a role in getting people to act on all issues that they care about and we care about. You can see it really tangibly in things like fundraising appeals. We know our fundraising appeals go miles better when people are angry and that there is a key moment for a person, that they are reacting. So, when you've got Tony Abbott as your Prime Minister, for however long he was Prime Minister, it was quite lucrative compared

> to more moderate government or political leaders that people perceive as sort of taking action, even if it's not that far or anything. You see that really materially. Absolutely. All about emotions. (Environmental NGO founder interview, November 2018)

Rather than becoming disinterested or withdrawing from action, people who want substantive action react more strongly to leadership that they perceive as opposing climate change interventions. Cyborg activism is, in this instance, strengthened in opposition to a clear risk to sustainability (Asenbaum 2018). At the same time, it can be good for environmental NGOs to have an explicitly conservative government in office—not in terms of the political decisions being made but in relation to how much easier it is to get community support in opposition to these decisions. Like any discursive field, counter-discourses to mobilising climate action have emerged as well, as social media has enabled ideological scepticism of climate change science to gain footholds in the United Kingdom and elsewhere (Lockwood 2010).

Tony Abbott's Prime Ministership included a return to John Howard's (1996–2007) climate policies, including that taking climate action was a threat to the national interest and 'essentially a sucker's game' (Eckersley 2013, 390). Previous to Abbott's leadership, Julia Gillard had become the first female Prime Minister and introduced a Carbon Tax that was intended to internalise carbon emissions in business and institutional costing (Curtin 2015). Abandoning the Carbon Tax that the Gillard government had installed was a key plank of Abbott's return to Howard's climate policies, as was abolishing funding to the Climate Commission. What is less well known of Abbott's leadership is the turn to effective action that occurred at the grassroots level, inspired by his anti-climate action stance. Climate change science and action has been politicised in Australia, as most resource management issues are, and traditionally those that identify with conservative politics do not support climate change action and mitigation, while those that are more progressive do (Unsworth and Fielding 2014). However, Eckersley (2013) points out that the difference between the two main political parties' policies on climate change in Australia is less significant than this pattern suggests, and the Greens, an environmental political group, are the only ones with substantive policies that might shift the climate action terrain.

Returning to the question of how emotions are involved in effective digital action, the way organisations grow and shift over time to respond to increasingly urgent climate change knowledge and persistently lax governance strategies is noteworthy. Leaders of these groups have a strong role to play in determining these changes, but so do their supporters. In her landmark book *Cyberprotest*, Pickerill (2003, 24) observed how 'Movement Intellectuals' are crucial for determining the affectual, cognitive and material scope of mobilising people and that their role could change: 'Such intellectuals could advocate new ways of viewing and utilising technology—to those within the movement and the general public—ways that might contradict with commercial desires for the technology'. The rise of social media driven activism has seen this bottom-up push emerge, along with the growth in sophistication of particular groups. For the Climate Council, the shift to independent funding and an increase in support staff with deep skill sets in different forms of media engagement has changed the way they approach communicating about climate change. Councillor for the Climate Council Lesley Hughes reports how this shift has occurred:

> *Lesley*: The Climate Council started with five of the original six Climate Commissioners plus Amanda McKenzie [the Climate Council's now CEO], with a bit of volunteer support. As donations have come in from the public we've been able to build our staff, and thus expand our activities. We have been fortunate to attract a lot of passionate and highly skilled (mostly young) people in both traditional and social media. With this team we have greatly expanded the communication activities and reach into the Australian community, as well as to business and all levels of government.

Lesley Hughes' academic work in the biological sciences rarely includes writing or researching subjectivities in a reflexive way. What is noteworthy about this comment is how the division of roles has grown in importance as the Climate Council has developed as an organisation. Similarly, Maz shares how in the early days of the environmental NGO that she founded, she was responsible for the social media communications aspect of the group and that this has become a task, in and of itself, for another member of staff. The separation of tasks to different employees and/or volunteers reflects specialised skills that are needed for effectively engaging with this mode of mobilisation.

## Strengths and Weaknesses of Digital Campaigning

Digital spaces are helpful in terms of building community, sharing information, gathering financial resources and obtaining results, but they are also no utopia for environmental activists and those seeking climate change action. The externally imposed constraints of digital spaces are significant and can starkly mediate action, as well as determine the distribution of resources within environmental NGOs. If someone is required to fill a defined role to manage social media engagement, then this draws down on the sometimes-stretched resources of an organisation. In addition, community and attention fatigue (Zuckerman 2013) can occur, where supporter bases tire of helping or even reading emails and social media feed. Corporate entities such as Facebook and Twitter also impose constraints as they morph to maximise profits. The Community Engagement officer for one peak environmental NGO described the costs of digital action in the following way:

> I think there's a potential for community fatigue, in terms of if you ask the base too much, too frequently, you can burn out your community in the list. I think there's also a lot of different voices competing in the same space, all for particular demographics, in particular the younger demographics are receiving a lot more information through digital channels. There's potential for messages to be missed or misinterpreted. It's striking that balance between content delivery, such as emails… It keeps changing, and it's something that we need to keep adapting to.
>
> I think there's also a reliance on platforms that can change or be filled up with other organizations doing the same thing. There's a definite risk.

> Third party petition sites that we used to use, you can no longer use in the same way, or they've now started to monetize things in different ways. Social media platforms change as well, so having to revamp can be costly. We need to do that quite frequently, in shorter and shorter cycles.
>
> Sometimes we can then also run into dead ends when we try and seek out new ways of doing it, as well, trying to find new avenues to receive the same results as what you used to be able to do. (Community Engagement officer interview, October 2018)

The work involved in 'keeping up' with digital changes to activate supporters is unending, either driven by an internally motivated interest in maximising the efficacy of sharing information, or externally imposed changes that produce 'dead-ends'. Environmental NGOs need a mixture, then, of proactive and reactive approaches to engaging with publics, warranting sophisticated management of this area of mobilisation. The particular forms of climate action that people in Australia who supported the creation of the Climate Council include diverse digital and non-digital modes, as Fig. 6.2 shows.

Navigating emotional terrains in activist communities adds a layer of complexity to the publics that activists and agents of change are trying to engage. People who are not involved and have no knowledge of an issue are characterised by Hallahan (2000) as 'non-publics' in his typology of publics and how they are activated. I think that the more-than-real is providing avenues for the non-publics group to engage differently now, with better access to information for many people (although this is geographically uneven) and with possibilities for different forms of political participation. Now, people and institutions can pursue social, cultural, environmental and political changes in other ways than rallying in public spaces; the more-than-real is a productive site for change seekers. So rather than framing these as 'non-publics', in the more-than-real this group could be framed as 'latent publics', or cyborg activists.

Sophisticated approaches to digital engagement are common to environmental activists that contributed to this study. Fundraising appeals, for example, are more effective over email, which is perceived as a private, direct communication method, while communicating

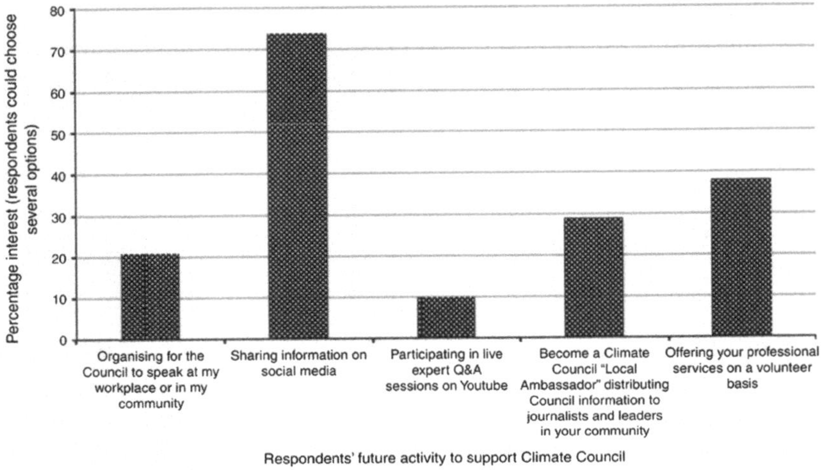


**Fig. 6.2** The climate action that supporters of the Climate Council are interested in pursuing (originally published in McLean and Fuller 2016)

messages to the general public needs social media such as Facebook or Twitter. When reflecting on the diverse approaches required for digital engagement, Maz, Environmental NGO founder, reported the following:

> Something like fundraising, you've probably noticed that you don't really get any fundraising appeals through Facebook much because it just hasn't, nobody's really been able to make that work well. Email, with a list to be able to fundraise appeals through, is far more effective than any other digital channel. Something like Twitter – it's an area we haven't done. We have a Twitter presence but we don't use it a huge amount. It takes a lot of time to engage in an online community like Twitter. It's the nature of how it works. It's good for influencers and people like journalists and thought leaders and so on, but it's not good for me; I think it's really for the public. For example, you probably didn't notice Mike Cannon-Brookes, head of Atlassian, was getting all fired up last night about renewable energy and stock prices so my environmental NGO started sending out some tweets to him about that to alert him to some work that we've done on how you get to 100% renewable, so it's good for something

> like that, but not reaching the masses. (Environmental NGO founder interview, October 2018)

Mike Cannon-Brookes did notice the environmental NGO's tweets and he tweeted the rich data that they provided him on how Australia could easily reach higher levels of renewable energy investment. The 'Fair Dinkum Power' push that introduced this chapter was connected to grassroots activism that has been working to materially transform energy consumption in Australia for ten years.

Some of the strengths of using digital technologies include easy and rapid communication, having the capacity to deliver messages to a desired audience, and building communities that might lead to societal and cultural transformations. As Table 6.1 summarises, different digital technologies are used for specific desired responses.

Organisations are rational about what digital spaces are worth their time and effort engaging with and strategically select areas to build their presences. At other times, ad hoc and opportunistic engagement can deliver useful interventions, such as the Fair Dinkum moment described above. The distinction between more private and more public spaces is of note here: environmental NGOs tend to view email as private and appropriate for fundraising appeals while social media does not usually garner such appeals. However, this may only be the case for already existing organisations and campaigns, rather than all of these efforts, as crowdfunding initiatives for novel interventions *do* gather momentum through social media.

Some see crowdfunding as an extension of neoliberalism. For example, Rodan et al. (2017) describe crowdfunding as fundamentally neoliberal in that it facilitates a user pays system; if you want a particular outcome, you put money directly towards that goal in crowdfunding processes. The label of crowdfunding as neoliberal is further concretised in how accumulations of payments by individuals removes the responsibility of delivering a particular service or thing from the shoulders of government. However, the case they give as an example of a neoliberalised institution is the Climate Council and there is an important distinction that is overlooked here. It is clear that because of its transformation, the Australian government no longer funds the Climate Council, however,

**Table 6.1** Activities undertaken by environmental NGOs in digital spaces and their possible outcomes

| Organisation effort | Desired response from supporters | Possible outcomes |
|---|---|---|
| Email | Funds for particular campaign or ongoing activism | Sustainability of organisation |
| Twitter | Retweet, comment, 'heart' | Greater awareness of organisation and issue; signup to email list |
| Facebook | Choose emoji reaction, comment, share | Growth of network; Greater awareness of organisation and issue |
| Outline steps for making public submission | Make a quality public submission | Official government inquiry; change in government policy |
| Lead a petition | Sign on to petition, share to networks | Contingent on goals of petition—may include cancelling an event, institutions adopting 100% renewable energy or stopping particular resource management projects |

the outcome of the crowdfunding was an institution that communicates climate change information, including risks of current and future impacts. All Australians benefit from this information and the ongoing existence of the Council as all Australians are experiencing the impacts of climate changes, to varying degrees. The argument that crowdfunding always delivers a neoliberal outcome does not hold up in this instance. Further, the Council is a not-for-profit—a far from neoliberal institutional structure and somewhere closer to a challenge to economic rationalism (McLean 2016). Perhaps crowdfunding could be positioned as a hybrid space, bringing together collective efforts to change social and environmental circumstances, but drawing on individualised funds and commitments to achieve these.

Of course, crowdfunding does have problems too as raising funds collectively does deliver income for profit-seeking crowdfunding platforms and this could be read as facilitating some neoliberal institutions. Other problems include the uneven support for crowdfunding initiatives—only popular issues and campaigns receive support. The biases and partiality of crowdfunding (Greenberg and Mollick 2015) lead Rodan et al. (2017) to suggest that crowdfunding facilitates a business-as-usual approach to social and environmental problems rather than a system-wide transformation. As I suggest above, there is, however, an element of hybridity and contradiction in crowdfunding that has also been put forward by others including Gehring (2016, no page numbers) who states that 'Crowdfunding represents an ideological mechanism simultaneously more transparent and obscured than other social networking sites such as Facebook or Twitter'. The transparency comes from crowdfunding's call out for involvement, the identification of the problem and the solution that is open for all to contribute to while the obscurity comes from the platform itself and the implementation of the solution thereafter.

Benefits of digital action are scaled according to the particular form of action undertaken. While digital technologies may have international reach, the impacts of environmental activism via these are frequently very grounded. For example, a recent successful community-building campaign was a petition by a peak climate change NGO to show support to the directors who lead the organisation. Rather than a standard petition to ask for an intervention or make a complaint, this petition aimed to bolster community engagement and encourage responsible environmental leadership. As Clara from the peak climate change NGO that organised the petition reports:

> Some of the actions are localized, like that petition. Others are targeted at a state or federal angle. Our organization also benefits greatly from these actions, so whether it's educating our audience in conversations about climate change through social media, to reaching new audiences in addition to our advertising, to then mobilize supporters and help them to engage with us. I think there's lots of different ways that we can benefit.

> It can also be simply like a tool for acquisition for us, in terms of the petition also, even though it had great positive benefits for the organization. It was certainly an acquisition tool for us, in getting 20,000 new emails.

> There's definitely dual purposes in terms of the way that that ran. I think, ultimately, all of our online action will benefit the climate in the end. Also, in terms of creating a safer, cleaner, more sustainable Australia. (Community Engagement officer interview, October 2018)

## Preparing for the Unexpected

The creation of social movements or moments of disruption in digital spaces is afforded and constrained by the particular dynamics of the medium used. For instance, Bruns and Burgess (2011) discuss the contingency of 'ad hoc publics' within Twitter as one portion of the public sphere, with communities of varying degrees of connection and affiliation. Some groups are strongly affiliated as followed and 'followees' while others are loosely connected by hashtag use and form ad hoc publics, distinct to issue publics that are more clearly demarcated. The contingency of ad hoc publics becomes clear in the hashtag instance as Twitter users may choose to use hashtags for emphasis rather than to explicitly contribute to broader conversations or debates in the absence of italics or bolding (Bruns and Burgess 2011). Further, ad hoc publics emerge and fade away as the crisis, issue or event is resolved. Bruns and Burgess (2011) give the example of the June 2010 deposing of then Prime Minister Kevin Rudd in an unexpected and successful #spill as an instance of an ad hoc public that grew over time and involved key Twitter users directing much of the conversation; journalists and politicians dominated the network for this #spill.

However, the temporal dimensions of ad hoc publics does not define all formation of publics, and far from all publics are ad hoc. Bruns and Burgess (2011) point out that publics may form following an event or in anticipation of a particular intervention. The complex contingencies of digitally mediated social movements become clear as the dimensions of dialogue are unpacked. From my conversations with environmental NGOs who are drawing on, and working with, digital spaces for social, political and environmental changes, a mixture of proactive and reactive strategies are drawn on to build momentum. In some ways, this mix of approaches is not necessarily new but the sophistication of approaches that are required to effectively engage is noteworthy and the possibilities and constraints of such processes are quite polarised. As mentioned earlier in this chapter, many environmental NGOs in Australia have separate roles for social media management and this has shifted significantly in the last five years from a situation where social media was once an adjunct role (Environmental NGO founder pers comm, October 2018). Contributing to, and engaging with, ad hoc publics is one task of such work while maintaining and extending existing networks of support forms a substantial other task.

The specialisation of communication roles also changes the public perception of organisations—partly as a result of changes in the way that these organisations themselves work. The Climate Council has morphed over the five years of their existence from an apolitical organisation communicating information on climate change to a group who are drawing on, and generating action from, the emotions and affect circulating around climate change realities. While Bruns and Burgess (2011) outline the differences between ad hoc and praeter hoc publics, the way publics form can involve hybrid strategies as well. The genesis of some institutions may not follow one straight narrative. The Climate Council emerged from crowdfunding efforts and Twitter outrage and these spontaneous aspects mixed with careful strategising by its leaders:

*Lesley*: The CEO of the Climate Council, Amanda McKenzie, had been the Communications Advisor for the Climate Commission and had very broad experience in advocacy, having previously helped establish the Australian Youth Climate Coalition. Just before the 2013 election she resigned her job in the public service and put together the business plan for the Climate Council. Once the Council was established, she became the CEO and has been the chief strategist since that the Council's inception. She and Tim Flannery have also been very active in raising funds from philanthropists.

The success of the early fundraising took us all by surprise. Many of the donations were quite small, but there were many thousands of them, so it all added up. Looking back I think that we served as a touchstone for people who were not only dismayed by the election of the Abbott government, but angry and frustrated that abolishing the

> Climate Commission was the first act of the government. Donating to the Council was a way for people to express their frustration in a positive and constructive way.

Coping with disappointment and the ensuing rage provided an opportunity for the emergence of a differently defined institution (McLean 2016). But the Commissioners (now Councillors) and administrative staff were also aware that their funding was about to be cut and made proactive efforts to transform the institution so they could continue the work that they were doing. This is an example of a hybrid moment of change where opportunistic community-building arose hand-in-hand with careful planning: or, in Bruns and Burgess' (2011) terms, the ad hoc mixed with the praeter hoc. The other elements of surprise come from changes introduced by external actors including technology corporations that own and manage these tools, governments that may intervene with new regulations and laws that affect the way digital spaces work, and competing groups seeking opposite goals. In Australia, the retention of metadata by the Federal government has prompted the emergence of new civil society groups such as Digital Rights Watch. The frequent shifting of algorithms and affordances of digital technologies by large corporations forces those who use these domains to react effectively and reassess whether continued presence is warranted and/or feasible. In response to a question about the social media strategy of Australia's peak climate change NGO, the Community Engagement officer offered the following:

> I think, definitely one of the things that characterises it is the need to be nimble and respond rapidly as the social media landscape keeps shifting. For example, like with the significant changes to Facebook's algorithm, that forced us to change the way that we use Facebook and the type of content that we produce. Our social media strategy, I think, primarily focuses on celebrating with new stories on renewables and solutions and creating that positive content. While we do have a couple of other types of stories, our primary focus is on these good news stories and creating shareable content that people will pass onto their friends. I think our social media advertising strategies overall aims to inform or capture new audiences, often repurposing our organic contents, such as videos, to get more out of the organic content that we produce for our channels. (Community Engagement officer interview, October 2018)

There are two main forms of strategy at play for this environmental NGO. First, responding to conditions imposed by external parties is paramount to ensuring ongoing effective engagement. It simply is not an option for the organisation to ignore dramatic digital shifts if they want to continue to use such spaces. Second, the organisation prioritises a positive communication approach in its work. Rather than reiterating the dire consequences of already happening climate change, the discursive approach of this organisation is to offer information on constructive alternatives to fossil fuel and encourage action through that information sharing process.

## Conclusion

Hybrid publics, forged by intertwined ad hoc interventions and strategic formulations by sophisticated organisations, form climate change action in the more-than-real. At various times, environmental NGOs build or respond to groundswells of affect while at others, reacting to top-down interventions by corporations and governments. The affordances of the more-than-real necessitate such reflexivity and multiple strategising. Always being ready, always trying to predict and/or react appropriately to opportunities or events is a key component of more-than-real environments. The preparedness to be surprised and then quickly act to maximise effective action emerged as a key theme in this research.

The way emotions around climate change are understood—and engaged with—is increasingly core to the work of environmental NGOs. Translating and unpacking outrage is part of this but certainly not the whole story. As conversations with those involved in leading organisations seeking climate change action show, and others charged with continuing community-building, reading nuanced emotional landscapes of environmentalism has become an important part of their work. Duffy et al. (2018) look to how local environmental leaders are working towards sustainability in a nuanced analysis of narratives of sustainability. The passions that drive such engagement are the focus of their study of local sustainability leadership. In this age of ongoing environmental crisis, Duffy et al. argue that engaging with Anthropocene challenges involves greater recognition of the role played by emotional and affectual responses that are involved in place-based connections. Further, the way Nature 2.0 emerges in this case, where social media is a key tool for activists, individuals and communities to achieve political and environmental outcomes is an important aspect of this transformation. Discourses and materialities of solar power in Australia share commonalities with other environmental-digital processes, such as how Hawkins and Silver (2017) trace discourses and materiality in their discussion of #selfies and #sealfies in Canada. The interactions of celebrity power with activism and local people's narratives of seal use show how power relations are messy in the more-than-real: claims to ethical human–animal encounters were misrepresented in the contestation of seal–human relations, and appropriated for specific activist and celebrity-endorsed goals. Similarly, the discursive and material aspects of solar power advocacy in Australia involves contestations at different scales and complex negotiations of power relations.

Last, an openness to transformation is an important part of current climate action in the more-than-real. The narrative that opened this chapter offered a glimpse into the curious dynamics currently at play in corporate, NGO, government and individual efforts to wrangle more effective climate change action. The fact that many environmental NGOs and a digital technology billionaire are more proactive than Australia's Prime Minister on a political and environmental issue that affects all Australians speaks volumes of the state of climate change action in this country. It also suggests that there are ruptures in the political landscape that are facilitated by more-than-real interactions that may transform the possibilities of climate change action as well as what activism means in the Anthropocene. The weird solidarities that we glimpsed in Chapter 2 are at play in climate change action within digital geographies as well. While Australia continues to have lacklustre national political responses to climate change, there is leadership from alternate actors, achieving material changes that are attempting to fill those structural voids. The more-than-real is providing avenues for climate change action groups to emerge differently now, with better access to information for many people, although this is geographically uneven, and methods of involvement are diversifying.

## References


Asenbaum, H. (2018). Cyborg Activism: Exploring the Reconfigurations of Democratic Subjectivity in Anonymous. *New Media & Society, 20*(4), 1543–1563.

Bates, O., Lord, C., Knowles, B., Friday, A., Clear, A., & Hazas, M. (2015). Exploring (Un)sustainable Growth of Digital Technologies in the Home. In *Proceedings of the Third International Conference on ICT for Sustainability* (ICT4S). Atlantis Press. Retrieved July 20, 2018, from https://www.atlantis-press.com/proceedings/ict4s-env-15/25836178.

Berkhout, F., & Hertin, J. (2004). De-materialising and Re-materialising: Digital Technologies and the Environment. *Futures, 36*(8), 903–920.

Brabazon, T., Redhead, S., & Chivaura, R. (2018). Trump Studies: The Double Refusal and Silent Majorities in Theoretical Times. *Cultural Studies Review, 24*(2), 3–25. https://doi.org/10.5130/csr.v24i2.5628.

Bruns, A., & Burgess, J. (2011). The Use of Twitter Hashtags in the Formation of Ad hoc Publics. In *Proceedings of the 6th European Consortium for Political Research (ECPR) General Conference 2011, University of Iceland, Reykjavik*. Retrieved July 20, 2018, from https://eprints.qut.edu.au/46515/.

Curtin, J. (2015). The Prime Ministership of Julia Gillard. *Australian Journal of Political Science, 50*(1), 190–204.

Duffy, M., Gallagher, M., & Waitt, G. (2018). Emotional and Affective Geographies of Sustainable Community Leadership: A Visceral Approach. *Geoforum*. https://doi.org/10.1016/j.geoforum.2018.09.005.

Eckersley, R. (2013). Poles Apart? The Social Construction of Responsibility for Climate Change in Australia and Norway. *Australian Journal of Politics & History, 59*(3), 382–396.

Fair Dinkum Power. (2018). *Fair Dinkum Power*. Retrieved March 31, 2018, from https://www.fairdinkumpower.com.au/.

Gehring, D. (2016, October). The Depoliticized Politics of Crowdfunding: A Critical Examination of the Darren Wilson Crowdfunding Campaign. *First Monday, 21*(10-3). https://doi.org/10.5210/fm.v21i10.6957.

Greenberg, J., & Mollick, E. (2015). Leaning In or Leaning On? Gender, Homophily, and Activism in Crowdfunding. *Academy of Management Proceedings*. http://dx.doi.org/10.5465/AMBPP.2015.18365abstract.

Hamilton, C. (2017, February 15). *That Lump of Coal*. The Conversation. Retrieved March 10, 2019, from https://theconversation.com/that-lump-of-coal-73046.

Hallahan, K. (2000). Inactive Publics: The Forgotten Publics in Public Relations. *Public Relations Review, 26,* 499–515.
Hannam, P. (2018, December 3). 'Six Panels a Minute': Two Million Australian Homes Now Have Solar. *Sydney Morning Herald.* Retrieved July 20, 2018, from https://www.smh.com.au/business/consumer-affairs/six-panels-a-minute-two-million-australian-homes-now-have-solar-20181203-p50jtn.html.
Hawkins, R., & Silver, J. J. (2017). From Selfie to #sealfie: Nature 2.0 and the Digital Cultural Politics of an Internationally Contested Resource. *Geoforum, 79,* 114–123.
Head, L. (2016). *Grief and Hope in the Anthropocene.* London: Routledge.
Hunt, E. (2017). Elon Musk: I Can Fix South Australia Power Network in 100 Days or It's Free. *The Guardian.* Retrieved May 1, 2019, from https://www.theguardian.com/technology/2017/mar/10/elon-musk-i-can-fix-south-australia-power-network-in-100-days-or-its-free.
Jasper, J. M. (1998). The Emotions of Protest: Affective and Reactive Emotions in and Around Social Movements. *Sociological Forum, 13,* 397. https://doi.org/10.1023/A:1022175308081.
Kim, Y. M. (2012). The Shifting Sands of Citizenship: Toward a Model of the Citizenry in Life Politics. *The Annals of the American Academy of Political and Social Science, 644,* 147–158.
Klein, A. (2018). Virtual Power to the People: A New Way of Generating Energy by Letting Communities Pool Together Is Getting Its First Big Test in South Australia. *New Scientist, 237*(3168), 22–23.
Kleine, D. (2019). Development. In J. Ash, R. Kitchin, & A. Leszczynski (Eds.), *Digital Geographies* (pp. 225–237). London: Sage.
Lockwood, A. (2010). Seeding Doubt: How Sceptics Have Used New Media to Delay Action on Climate Change. *Geopolitics, History and International Relations, 2*(2), 136–164.
Lockwood, M. (2018). Right-Wing Populism and the Climate Change Agenda: Exploring the Linkages. *Environmental Politics, 27*(4), 712–732. https://doi.org/10.1080/09644016.2018.145841.
Macquarie Dictionary Publishers. (2018). *Fair Dinkum.* https://www-macquariedictionary-com-au.simsrad.net.ocs.mq.edu.au/features/word/search/?word=fair+dinkum&search_word_type=Dictionary.
McLean, J. E. (2016). The Contingency of Change in the Anthropocene: More-Than-Real Renegotiation of Power Relations in Climate Change Institutional Transformation in Australia. *Environment and Planning D: Society and Space, 34*(3), 508–527. https://doi.org/10.1177/0263775815618963.

McLean, J. E., & Fuller, S. (2016). Action with(out) Activism: Understanding Digital Climate Change Action. *International Journal of Sociology and Social Policy, 36*(9/10), 578–595.

Morrison, S. (2018). *'You Asked. Here's My Answer' Scott Morrison*. Retrieved November 5, 2018, from https://www.facebook.com/scottmorrison4cook/videos/496591834153066/.

Penney, J. (2015). Responding to Offending Images in the Digital Age: Censorious and Satirical Discourses in LGBT Media Activism. *Communication, Culture & Critique, 8*(2), 217–234.

Pickerill, J. (2003). *Cyberprotest: Environmental Activism Online*. Manchester: Manchester University Press.

Rodan, D., Mummery, J., & Henkel, C. (2017). The Charity Model Is Broken: Crowdfunding as a Way to Democratise, Diversify and Grow Funding for Social Change? In F. Martin (Ed.), *Refereed Proceedings of the Australian and New Zealand Communication Association Conference 2017—Communication Worlds: Access, Voice, Diversity, Engagement*. Retrieved July 1, 2018, from https://anzca.net/index.php?option=com_content&view=article&id=247:2017-conf-p2&catid=94:past-conf-p2.

Sandover, R., Kinsley, S., & Hinchliffe, S. (2018). A Very Public Cull—The Anatomy of an Online Issue Public. *Geoforum, 97*, 106–118.

Spring, V., Cameron, C., & Cikara, M. (2018). The Upside of Outrage. *Trends in Cognitive Sciences, 22*(12), 1067–1069.

Unsworth, K., & Fielding, K. (2014). It's Political: How the Salience of One's Political Identity Changes Climate Change Beliefs and Policy Support. *Global Environmental Change, 27*, 131–137.

Zuckerman, E. (2013). *Rewire: Digital Cosmopolitans in an Age of Connection*. New York: W. W. Norton.

# 7

# Delivering Green Digital Geographies? More-Than-Real Corporate Sustainability and Digital Technologies

Part of the allure of digital technologies has been the purported increase in efficiency and sustainability that they offer. The discourses of paperless offices and greater digital efficiencies have helped to produce the gloss of the more-than-real. The packaging of digital tools has included promises of a turn from resource-intensive work and lives, towards automated and effective processes. However, are the large-scale corporations that facilitate digital technologies, and their spread throughout the world, delivering on these promises? To gain a deeper understanding of the possibilities and constraints of digital spaces, we need to evaluate the ways in which these tech companies operate, and continue to interrogate claims of sustainability. This chapter examines how environmentally ethical these institutions are and what 'green' and equitable credentials they claim.

An analysis of the sustainability of digital technologies can occur at multiple scales, from the global to the individual. I approached the leading technological companies in Australia who are involved in key parts of the digital ecosystem—from Internet Service Providers to cloud providers to tech corporations, including Amazon, Rodin, EMC, Telstra, Optus, Servers Australia, Apple and Microsoft. None of these


J. McLean, *Changing Digital Geographies*,
https://doi.org/10.1007/978-3-030-28307-0_7

companies were interested in, or available for, an interview and so I draw on publicly available information, including corporate policy statements on sustainability and media coverage of their sustainability initiatives. The information presented in this chapter comes from the analysis of published annual reports, sustainability reports and critiques of tech company's green credentials. The larger corporations tend to report environmental and social equity issues in more detail than the smaller corporations in Australia but as these disclosures are self-reporting, warrant some scepticism about their claims. Further, even in the best-case scenario that the data is accurate, the high demand for Internet Service Provision is producing increasing carbon emissions. I was also able to interview an Information Systems and Business Analysis expert at Macquarie University, Professor Deborah Richards, who has been teaching students about sustainability initiatives, or the lack thereof, for over ten years. Her insights show how much work is still needed in this area.

The digital ecosystem that forms the focus of this chapter is only partially glimpsed in everyday life due to the complexities of the interactions between the different sectors and portions of communication systems; wi-fi helps produce the increasing invisibility of these digital ecosystems. The social, political and cultural implications of how digital geographies are made are similarly complex; transparency in energy consumption and resource needs eludes, and sources of essential equipment are frequently not made visible. To unpack these complexities, I draw on Plumwood's (2008) 'shadow places' notion that offers a way of reading externalities and displacements of responsibilities across multiple scales, between the Global South and North. Plumwood (2008) argues that the Global North benefits from the distancing and obscuring of key aspects of commodity chains and that inequality is underpinned by this marginalisation of the Global South. The idea of shadow places serves to highlight the way that power intersects with cultural practices and makes visible often hidden neoliberal and colonial practices (McLean et al. 2018). Plumwood asks us to look into the shadows and reveal that which is hidden. Shadow places in this context are somewhat different to the digital shadows theorised by Graham (2013) who argues that cities are made by their layers of digital content generated about them, as much as by the buildings, roads and people that form them. This chapter offers an analysis of a segment of Australian digital technologies but it is relevant beyond this context in that the capitalist systems that underpin these digital technologies are mirrored in similar ways around the world. For brevity, I focus on telecommunication companies and data server companies rather than the whole digital ecosystem.

## Digital and Sustainable Tech Companies?

Sustainability in digital technologies is often framed as a problem that smart design can solve. Such reasoning posits that if people can see and measure their energy consumption, then they will reduce the level of their energy use and make their lives more sustainable, eventually saving the planet in some distant utopic time and space. The assumptions underlying this claim are unpacked by Strengers (2013, 36) in her skewering of the Resource Man:

> In his ultimate imagined state, Resource Man is interested in his own energy data, understands it, and wants to use it to change the way he uses this resource.

The problems with this ideal person are multiple, according to Strengers, and include the complete absence of lived experience and a capacity to talk to diverse ways of being. I hold this salient critique of Resource Man in mind as I detail efforts by corporations to address sustainability dilemmas.

Research on the carbon footprint of digital technologies is emerging that shows how far from sustainable our digital lives already are. A case in point is the carbon emissions alone that accompany the life-cycle of AI technology: Strubell et al. (2019) found that training a large AI model emits 315 times more carbon than a round the world flight. The discourses and practices of sustainable development have grown since the Brundtland Report of 1987. The possibilities and constraints that sustainable development discourse offers to digital technologies are substantial, as Floridi (2001, 4) points out:

> Sustainable development means that our interest in the sound construction of the infosphere [synonymous with digital technologies] must be associated with an equally important, ethical concern for the way in which the latter affects and interacts with the physical environment, the biosphere and human life in general, both positively (e.g. telework as a solution for traffic and fuel pollution) and negatively (e.g. rising energy consumption, ICT generated waste, computer-related forms of illness).

Digital technologies have been facilitated by institutions that laud themselves as sustainable corporations, using arguments such as reduction of damage to environments by cutting down on paper use and reducing the need for travel. It is worthwhile, then, to dissect the sustainability discourses that are currently dominant in major telecommunication companies around Australia to examine the features of these sorts of claims. Table 7.1 shows different key features of claims to sustainability and their measures by the biggest Australian telecommunication companies.

Energy consumption, as indicated by carbon emissions, is increasing across the board while social inclusion measures, such as gender equity, do not seem to be shifting. The choice of what aspects of equity are included in these corporation's measures is interesting. Gender equity measures are a part of the reporting process but other key axes of difference, including age, disability, class and ethnicity/race, are not a part of the evaluations. The close association of gender equity with sustainability is put forward in nearly all of the available reports from tech companies that I reviewed. The Optus Sustainability Scorecard presents this visually and succinctly as captured in Fig. 7.1.

There are six measures of economic performance offered in the Optus scorecard, three of marketplace and customers, four of 'People', five relating to community and three for environment—all of which pertain to energy measures. The 'Community' is characterised by the amount of money invested in 'community', children 'reached' by cyber safety programmes and youth literacy. Reducing 'community' to a data point further obscures the complexity in communities that is simplified by the singular community label (Clark 1967). Optus connects digital literacy to sustainability as shown in Fig. 7.1. A thorough assessment of the

**Table 7.1** Sample of digital corporations in Australia and their claims of sustainability

| Company | Telstra | Optus | TPG | Dodo |
|---|---|---|---|---|
| Carbon emissions (trend of greenhouse gas emissions 2015–2018) | Decreased from 1,571,066 to 1,440,260 tonnes $CO_2$ | Increased from 396,863–401,982 tonnes $CO_2$ | No information in annual reports or sustainability report | No data on emissions |
| Energy consumption (gigajoules 2015–2018) | Decreased from 6,009,922 to 5,928,522 | Increased from 1,657,262 to 1,724,106 | No information in annual reports or online sustainability report | No data on energy consumption |
| Digital inclusion—community investment | Aiming for 18 AUD million investment by 2020, 6 million AUD so far | Up from 8.7 to 9.4 million AUD | No reported data | Have some community scholarships and sponsor the Telco Together Foundation |
| Gender pay equity | No parity (1:0.98) | No parity (1:0.96) | Not available | Introducing a gender pay gap check |
| Gender equality | Target of 32% women participation by 2018, only at 30% | Leadership: Increased from 8 to 22 executives | 20.7% women, 79.3% men.<br>No female directors | 2 out of 10 directors are women |

*Sources* https://www.optus.com.au/content/dam/optus/documents/about-us/sustainability/Sustainability-Scorecard-FY2018.pdf, accessed 15 October 2018

https://exchange.telstra.com.au/sustainability/data-downloads/

https://exchange.telstra.com.au/achieving-gender-equality-our-bias-for-action/

https://www.tpg.com.au/about/pdfs/2018%20Workplace%20Gender%20Equality%20Public%20Report.pdf. Public report TPG WGEA 2017–2018

https://www.smh.com.au/business/workplace/top-companies-failing-on-gender-pay-gap-face-investor-revolt-20180723-p4zt3d.html

TPG Sustainability report 2017, accessed 15 October 2018

https://www.tpg.com.au/about/pdfs/FY17SustainabilityReportv3.pdf

Dodo https://vocusgroup.com.au/media/1379/vocusannualreport2018.pdf

https://vocusgroup.com.au/media/1291/sustainability-report-2017.pdf

**Optus Sustainability Scorecard**

| Category | Indicator | 2017-18 | 2016-17 | 2015-16 |
|---|---|---|---|---|
| Economic Performance (GRI 201-1) | Economic value generated ($million) | 8,710 | 8,425 | 9,115 |
| | Operating costs (ex-staff costs, $million) | 4,930 | 4,672 | 5165^ |
| | Staff costs ($million) | 1,193 | 1,232 | 1277^ |
| | Dividends to shareholders ($million) | 350 | 350 | 350 |
| | Tax expense ($million) | 342 | 330 | 379 |
| | Economic value retained ($million) | 1,886 | 1,833 | 1,935 |
| Markeplace & Customers | 4G population coverage | 98.5% | 96% | 94% |
| | 4G base stations | 6,783 | 5,872 | 4,850 |
| | Telecommunications Industry Ombudsman complaints / 10,000 customers | 3.7 | 2.66 | 2.5 |
| People | Employee engagement score* | 77% | 72% | 76% |
| | Women in executive management | 22 | 18 | 8 |
| | Gender pay ratio (operations and support) | 1:096 | 1:098 | 1:096 |
| | Workplace injury frequency rate** | 2.8 | 0.8 | 0.8 |
| Community | Community investment ($million) | 9.4 | 8.2 | 8.7 |
| | Children reached by cybersafety programs | 59,023 | 55,140 | 51,728 |
| | Youth literacy rate improvement*** | 94% | 93% | 95% |
| | Employee volunteering hours **** | 12,128 | 16,472 | 16,194 |
| | Volunteering participation rate | 23% | 21% | 21% |
| Environment | Energy consumption (gigajoules) | 1,724,106 | 1,702,440 | 1,657,262 |
| | Direct & indirect greenhouse gas (GHG) emissions (tonnes $CO_2$-e) (Scope 1 and Scope 2) | 401,982 | 400,280 | 396,863 |
| | Carbon emissions intensity (tonnes $CO_2$-e/TB) | 0.17 | 0.22 | 0.31 |

**In 2016 we changed methdologies for measuring engagement. FY17 result onwards is per new methodology. Using previous methodology, our FY17 Engagement score would be 75%.*

***In FY2018, all injury or work related injuries submitted to Optus' insurer, irrespective of time lost, have been captured in reporting data. In FY2017, only 'accepted' claims with more than one day lost were included in Optus' workplace injury incidence rate. This difference in an improved reporting methodology has resulted in what appears to be an increase in the injury frequency rate this year, when in fact the change was, to align with the ILO.*

****Data from The Smith Family 2017 student2student Mobile Program Evaluation Report*

*****Volunteering hours in 2017-18 rdecreased due to a revised calculation method from previous years. Volunteer participation rate increased in 2017-18.*

*^ Numbers restated from those originally published to ensure comparability over time*

**Fig. 7.1** Screenshot of Optus Sustainability Scorecard (2018)

supply chain of their products and their sustainability is not included in the scorecard. Rather, Optus has chosen to define environmental sustainability as tied to just carbon emissions monitoring. Defining environmental sustainability in such a way is somewhat limiting and reductive and avoids a thorough assessment of the sustainability of the corporation.

Big tech companies are pursuing sustainability in terms of carbon dioxide emissions too. For instance, Apple claims to run on 100% renewable energy as of 2018 but, as Statt (2018) points out, this is not through 100% use of renewable energy in all parts of the commodity chain. Rather, the renewable energy claim is supported by carbon offsets. For example, in China where renewable energy is not available in some factory settings, renewable offsets are purchased by Apple to compensate for this fossil fuelled energy use. Offsets are not going to absolve the environmental harm produced by such energy use (Beder 2014) as the systems in place for producing offsets are not environmentally best practice. Beder (2014) argues that creating plantations as carbon sinks, a popular mode of producing carbon offsets, expands monocultural, environmentally unsustainable practices. Simply offsetting the soaring consumption of carbon for digital technologies will not reduce the climate impacts of such practices or the non-renewable energy continually consumed. Such attempts at fungible mediation, as Pasek (2019) might say, do not produce safer environmental futures. Fungible mediation is 'a set of semiotic strategies that serve to translate local relations into generic commodities that can be bought and sold at a distance, obfuscating the question of accountability in favor of the formal logics of accounting' (Pasek 2019, 2). In the case of Apple offsetting its carbon consumption in China, the local relations of coal use are attempted to be overridden by the investment in green energy and provide conscience soothing for troubled Apple consumers. Fungible mediation in that case involves offsetting unsustainable carbon emissions in shadow places with other activities, elsewhere, that assuages consumer guilt in the Global North.

## Discourses of Sustainable Corporations

The discourse of sustainability is deeply embedded in corporate talk around the world and is frequently not backed up by action. For example, research published last year by the Boston Consulting Group and the MIT demonstrated that '90% of executives see sustainability as important, but only 60% of companies have a sustainability strategy' (Kiron et al. 2017, 2). The likelihood of the sustainability strategy then translating into material change is possibly low too, given the frequent emphasis on profits rather than responsible environmental decision-making. Such deep inconsistency is not unusual within capitalist contexts where accountability for so-called 'externalities' such as the environment is low. Widespread support for sustainable technology exists yet, despite this, there is little indication that the practices of these institutions are achieving sustainability. For example, claims abound relating to the supposed sustainability of this corporate sphere, such as 'Tech companies are among the leaders of the sustainability movement, which is not surprising because they're in the business of innovation' (Fox 2011, no page numbers). The headline of that article, '5 Ways Tech Companies are Committing to Sustainability', gets to the core of the problem. These companies may be ‘committing’ to sustainable practices but are they achieving this poorly defined goal? Reilly and Hynan (2014) point out that the discursive work of asserting corporate sustainability provides a window in to the culture of particular organisations and more research is required to establish if corporations are actually being environmentally responsible.

Within Australia, the statistics on plans to achieve sustainability are not more encouraging. For instance, the non-government organisation Market Forces published a report in 2018 that analysed climate data, governance and action by 73 companies listed on the Australian Stock Exchange and found that most did not have clear established targets for reducing carbon emissions (58%). Further, only 16% of the companies that were examined by Market Forces had a clear plan to reduce carbon emissions (Market Forces 2018). It is important to note that none of these companies were technology industry leaders but they all do use digital technologies. If they have not even developed plans to reduce carbon emissions, then factoring in carbon emissions for digital technologies would logically not be a part of their planning. The gap between discourses of doing something about climate change and actually setting up plans with identified targets to reduce carbon emissions is worrying.

Given the gap between discourse and practice on sustainability, it is no surprise that the paradigms of sustainable development and sustainability have received deep critique since their popularisation. Banerjee (2003) argues that sustainable development is more about allowing economic growth to continue while offering tokenistic inclusion of social and environmental concerns. He sees the expansion of sustainable development ideology into development programmes in periphery countries as another arm of colonialism and Imperialism. In later work, Banerjee (2008) critiques corporate social responsibility and the internalisation of sustainability discourses to extend corporate reach. The particular ways in which indicators of sustainability are enthroned within corporate practice allow for a business-as-usual approach rather than a transformation of these institutions. Banerjee’s (2008) tracing of legal moves that enabled externalisation of environmental and social costs of corporations in the 1980s and onwards highlights the ways that core priorities of corporations have not transformed.

Some scholars are optimistic that the sustainability paradigm could deliver better ecological, social and economic outcomes. For instance, Wheeler (2016) argues that within urban planning, the sustainability paradigm offers a productive way to shift thinking about planning dilemmas in a more holistic way. Others use sustainability as a central framework for understanding the multiple pressures of changing urban contexts. For instance, McManus (2005, 200) argues that 'the future of Australia, as with much of the world, is about the development of sustainable cities'. He uses the language of vortex cities to explain the problematic qualities of urban Australia and argue for a transformation towards sustainable cities. Christensen et al. (2015) argue that the conceptual ambiguity that 'sustainability' carries is a potential strength. The openness of sustainability discourses can work to mobilise diffuse and divergent stakeholders in a loose coalition rather than channelling closed off views of what the concept means and how it should be deployed (Christensen et al. 2015). These differing views on sustainability share an interest in critically examining how the term is shifting, and what sustainability means in varying contexts, including the way that sustainability can be used to continue exploitative practices.

Environmental concerns in corporate contexts can receive a sort of greenwashing, a longstanding practice of corporate entities trying to produce a balanced 'triple bottom line', as dubious an effort as that may be (Cornelius et al. 2008). But drilling into the discourses of sustainability that are accompanying the extension of digital technologies into everyday lives shows insidious possibilities. The adoption of more devices, apps and software tools to accompany multiple ways of being in the world is often facilitated with these promises of more efficient households and businesses, and less environmental impact. The paradox emerges when we see that these technologies are not necessarily better for us, or the environment, and yet we absorb them into our lives with the assurances that they are.

Researchers and teachers in computing studies have attended to these paradoxical human–technology relations. For example, Professor Deborah Richards teaches Information Systems and Business Analysis in the Australian technology industry and brings these issues to light in curriculum from first to final year studies. One of the tensions that Deborah Richards emphasises in this curriculum relates to the ever-increasing desire for connectivity and the environmental costs of this desire:

> We have these insatiable desires for information, and to be connected... But people really need to think, because the people just want to stream, they want to have access, and, I mean, now everybody has unlimited downloads, but it all comes with a cost. And people aren't recognizing that cost. So there's more and more data centres. (Deborah Richards pers comm, October 2018)

Bringing these issues into conversation for students is one way in which the seeds of change might be planted. Rather than celebrating the always-new, always-possible world of digital technologies, recognising the limits of these technologies, and ultimately this planet, talks to the broader context of these 'insatiable desires'.

Frequently, even in critiques of greenwashing, the tools and infrastructure that support digital action are framed as almost separate to the corporations that are more obviously environmental offenders. Part of this mistaken framing may be due to our oversight of how 'infrastructures are embedded in our everyday practices and are multi-sited in nature' (Horst 2013, 148). Horst is referring to the qualities of infrastructures as multiply positioned. In the context of greenwashing, Delmas and Burbano (2011) give a nuanced account of corporate efforts to dress up their efforts as environmentally conscious and how governments, non-government organisations and individuals can challenge corporate efforts to obfuscate in terms of environmental damage, including how product-level greenwashing or firm-level greenwashing can occur. However, they also seem to place digital campaigning as positioned outside the net of possible greenwashing: 'Activists' and NGOs' access to consumers and the public has increased through use of Twitter and Facebook, YouTube campaign videos, and other internet based platforms... Green activists and environmentally oriented nonprofits on the lookout for greenwashing thus have an easy, inexpensive means to spread information about and campaign against greenwashing' (Delmas and Burbano 2011, 71).

Digital technologies are instrumental in this 'call-out' of corporate efforts to defraud consumers and it is important to also factor in the corporations that are facilitating such digital lives in terms of their impacts—especially social, environmental and cultural impacts.

Yes, digital technologies are relatively easy and inexpensive to use when compared to paper-based mail outs and phone calls but the externalisation of costs relating to these tools persist, and perhaps undermine such claims.

Another challenging dimension of sustainability and digital technologies is the way that social and cultural differences are addressed by corporations. There is extensive literature on gender and technology, and still the relationships between digital transformations and other axes of social difference are less well explored. For example, Scholz (2016) suggests that we need to understand the implications of expanding engagement with digital spaces under a subheading in a position paper on the topic that states 'The Social Dimension of the Digital Revolution Is Not Yet Well Understood'. Within this subheading are key problems associated with a better understanding of digital spaces including the singularity of 'the social dimension', an assumption that we will one day understand fully socio-technological relations, and the celebration of the digital revolution. Following Hickey-Moody and Wood's (2008) engagement with Second Life (a VR world) and the diverse cultures therein, it is important to note how social and cultural worlds are irreducibly complicated.

## Measuring Green IT

Interest in building green digital technologies does not seem to be growing at this point in time. The research that I have done for this book, including literature reviews, desktop analysis, interviews with experts, and attempted interviews with corporations, suggests that this fading interest may be associated with the escalating use of digital technologies and the challenges in measuring the impacts of such usage. Within Australia, the latest report on the sustainability of the Information and Computer Technology sector was produced by Fujitsu in 2014. The 'ICT Sustainability: Australian Benchmark 2014' report (Fujitsu 2014) gives a snapshot of the industry and practical tips for improving the sustainability of the sector. It was endorsed by the Worldwide Fund for Nature's CEO at the time and the then New South Wales Minister for the Environment, Rob Stokes, so it carries some not-for-profit and government legitimacy. The data in the report is clearly presented and shows that consumption of energy within ICT sectors is growing. Mike Foster, then CEO of Fujitsu Australia and New Zealand and now Head of the Oceania Region, wrote:

> Adding fuel to the fire is the growth of ICT. The proliferation of smart mobile devices, data and our incessant need to be connected 24/7 have resulted in the ICT industry undergoing exponential growth in recent years with no signs of slowing. It is now estimated that ICT as an industry is responsible for around 3% of the word's total global emissions. While this may seem insignificant, to put these figures in perspective, ICT emissions are comparable to those of the total global aviation industry or the total annual emissions of Australia, New Zealand and Indonesia combined. Further, given current growth rates, experts estimate total emissions will rise to 6% - or in some cases 10% by 2020. At 10% our emission's [sic] will be on par with the global aluminium smelting or cement production industries and India or Russia are added to the afore mentioned list. In short, ICT's impact on the environment is of growing concern. (Fujitsu 2014, 4)

Foster goes on to say that ICT needs to reduce, rather than increase, the carbon emissions that it produces, and 'play a pivotal role in driving down the remaining 97% of emissions through ICT enabling technology' (Fujitsu 2014, 4). To enable such 'driving down', the full accounting of emissions needs to occur, including comparing the amount of carbon emissions required to produce and maintain the 'ICT enabling technology' compared to that which it replaces.

The narratives offered by Fujitsu (2014) include celebrations of the innovations that digital technologies afford. For example, the simple swap of hefty manuals for iPads by aviation pilots is heralded (Fig. 7.2).

The undeniable—and easily measurable—fuel savings are important and a good contribution to mitigating carbon emissions. However, the whole picture of sustainability is obscured in this narrative as the environmental costs and benefits of the iPad *itself* is left out of the equation. The production costs, the maintenance costs (energy, repairs), and the lifecycle costs of the iPad are externalised, and only energy savings

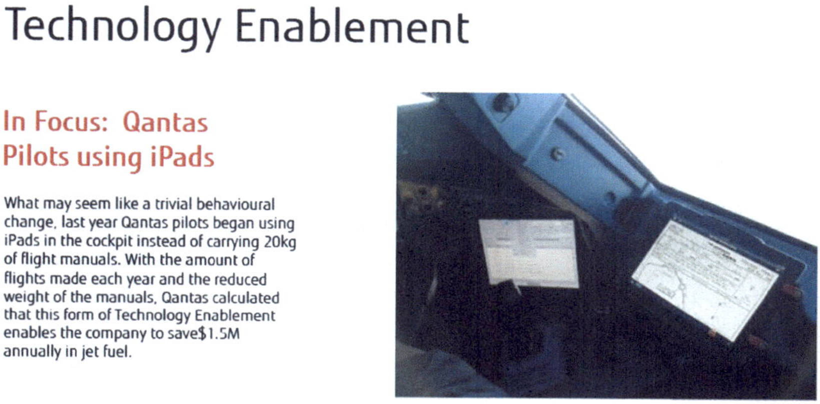


**Fig. 7.2** Digital technologies providing fuel savings—image from pdf of Fujitsu report (Fujitsu 2014, 18)

of replacing heavy paper-based texts is calculated. This is an example of the frequent flattening of the environmental costs and benefits of digital technologies. The assumptions underpinning the replacement value of an iPad for a stack of books are only apparent if readers query such data.

Deborah Richards notes that 'green IT' is not really high on anyone's agenda at the moment, if it ever was, as issues related to big data and automation are of broad and current interest.

> So I guess AI, and robots, and cyber security has taken over. I can't say that green IT was ever a really hot topic. I think it must have been in a little bit of a lull, and it was sort of, "Oh, yeah. Okay. Let's show a bit of interest in green IT." But I think it's probably off the agenda now and everybody's just looking at big data.

Despite innumerable policy agendas across the world addressing sustainability, from government to corporate institutions, emerging since the 'Limits to Growth' (Meadows et al. 1972) intervention, the shadow places that support the continued consumption of the Global North persist. The stark realities that might emerge from a full accounting for the many impacts of digital technologies might dull the shine they give to daily lives.

In attempting to increase awareness of green IT, one student at a time, Deborah Richards encourages students to consider the impacts of their choices in designing digital technologies. Rather than ignoring environmental costs and benefits, she asks students to factor these issues into designs:

> I invite students to ask the question, "Well, what is the footprint of this one versus the other?" So if they're building a website, thinking about, "Well, do I need to zip something up so that it's not such a large folder to send?" But then you need to think about, "Well, what is the effort involved in zipping it up and then unzipping at the other end?" So again, this is a complex issue to address… (Richards October 2018)

Market forces will probably not encourage the application of such thinking in corporate environments unless full assessment of environmental costs and benefits is regulated and promoted. However, supporting students to extend and develop their critical faculties to implement such initiatives is instrumental in making possible such transformations—and there are clear benefits of equipping students to do this.

The latest Fujitsu ICT Sustainability Australian Benchmark report came out in 2014 and the literature on this topic seems to taper off around then as well. Andreopoulou (2012) suggests that Green ICT can improve environmental outcomes as they increase efficiencies, can reduce consumption of certain resources and afford better access to essential digital tools in different contexts. However, it is the behaviour of consumers that may challenge such sustainability according to Andreopoulou (2012, 7): Green ICT 'cannot substitute people and their behaviour that still constitutes the most critical factor within environmental protection and sustainability'. While individual consumers are a part of the equation, it can not be framed as where all responsibility lies. As corporations are driving these ICT innovations, they must be at least partially held to account for spiralling consumption rates. Further, governments must play a regulatory role in the production of digital technologies to ensure that sustainability is fostered in design protocols and practices. As Bronk et al. (2010) calculated in an

**Table 7.2** Global carbon emissions associated with digital technologies—from Bronk et al. (2010)

| | 2009 | 2015 | 2020 |
|---|---|---|---|
| Data centres | 121.30 | 229.87 | 369.48 |
| PCs | 126.69 | 222.41 | 516.55 |
| Mobiles | 1.54 | 3.74 | 6.58 |
| Gaming consoles | 11.23 | 26.04 | 40.22 |
| Carbon conversation number | 1.3 | 1.265 | 1.23 |
| Total | 260.77 | 482.06 | 932.84 |

extensive assessment of carbon emissions and forecasted levels, these levels of consumption are increasing (Table 7.2).

The data that Bronk et al. (2010) capture is worrying and corroborated by other estimates of increasing data and energy use. For example, the United States Data Center Energy Usage Report (Shehabi et al. 2016) shows that 1.8% of total American electricity use is dedicated to the Internet and that consumption is growing rather than stabilising. Helman (2016) shares this research but globalises the data in the title of his Forbes piece 'It Takes 70 Billion Kilowatt Hours A Year To Run The Internet' and obscures the American specificity of this research. If it were more accurately reported, the data would be more useful, and potentially more alarming.

## Conclusions

This chapter has been one of the most disheartening to write and, I imagine, probably not that inspiring to read. The state of sustainability in digital technologies is not clear as data is partial, obfuscations abound and self-reporting dominates what evaluation of sustainability does exist. The best that is offered by mega corporations like Apple is a claim of 100% renewable energy use—but this is through offsets *and* purchase of renewable power rather than a transition to entirely renewable energy use. Shadow places continue to bear the costs of increasing consumption of modernist tools, including digital technologies. Digital geographies are formed in part by these attempts at fungible mediation (Pasek 2019) and when understood in context, these efforts to reduce environmental impacts are shallow and transparent. The data that is made publicly available by digital technologies companies, such as scorecards on sustainability measures, must be critically engaged with as they are self-produced and not independently audited. Corporate control over discourses relating to social-technological relations is not limited to claims to sustainability. As Sadowski and Bendor (2019) point out, corporations such as IBM and Cisco discursively dominate the smart city imaginary and breaking this stranglehold is important if alternative values and worldings are to come forth. Similarly, the discourses of sustainability produced by Apple et al. need puncturing, countering and reframing.

The literature on digital technologies and sustainability is grappling with the same questions that activists, academics and technologists were considering at the beginning of the Internet age. Floridi's (1999) comments on the infosphere include exhortations to consider digital spaces as real and containing polar possibilities embedded in an ethical frame. Ethical decision-making in digital technologies surrounding sustainability questions is often left up to individual corporations or consumers to regulate and manage—despite the opportunities that governments could take by expanding measures such as energy consumption information on devices and broadband usage. It seems that debates about Artificial Intelligence are taking prominence over these questions of sustainability in the digital technologies literature and mainstream discussions. The contradictory qualities of human–digital relations where environmental concerns are frequently foregrounded but then disregarded in calculating the sustainability of communities and institutions are inherent to the more-than-real. The infrastructure and energy needs to support environmental activism, for example, are often not incorporated into the accounting of the effectiveness of environmental NGOs. The contradictions of Nature 2.0 are visible here: Büscher (2017) recognises how the everyday things that seem inconsequential, routine and dull can significantly transform human–nature relationships. Engaging ethically with humans and non-humans, nearby and in shadow places, which the more-than-real pressures, is one way to start refiguring these relationships. More worrying contradictions appear when corporations like Amazon seemingly backtrack on 2014 commitments to 100% renewable targets and start forming relationships and partnerships with fossil fuel companies like BP and Shell (Merchant 2019). Such rapid shifts in the wake of inarguable evidence of accelerating climate change and other environmental dilemmas in the Anthropocene cannot be left unnoticed or unchallenged.

The groundwork on outlining the contradictions and limitations in digital-human technology imaginings and sustainability has been done by academics such as Strengers (2013) and Andreopoulou (2012). The 'Resource Man', the non-existent ideal smart energy consumer, that Strengers offers in her examination of smart city rhetoric resonates with the discursive critique of sustainability put forward by corporations as outlined in this chapter. The gendering of this identity is not accidental: corporations, especially digital ones, frequently reproduce masculinist thinking.

## References

Andreopoulou, Z. (2012). Green Informatics: ICT for Green and Sustainability. *Agrárinformatika/Journal of Agricultural Informatics, 3*(2), 1–8.

Banerjee, S. B. (2003). Who Sustains Whose Development? Sustainable Development and the Reinvention of Nature. *Organization Studies, 24*(1), 143–180.

Banerjee, S. B. (2008). Corporate Social Responsibility: The Good, the Bad and the Ugly. *Critical Sociology, 34*(1), 51–79.

Beder, S. (2014). Carbon Offsets Can Do More Harm Than Good. *The Conversation*. Retrieved January 12, 2019, from https://theconversation.com/carbon-offsets-can-do-more-environmental-harm-than-good-26593.

Bronk, C., Lingamneni, A., & Palem, K. (2010). *Innovation for Sustainability in Information and Communication Technologies (ICT)*. Internal Report, Rice University. Retrieved January 10, 2019, from https://pdfs.semanticscholar.org/8bbb/43e5dc9562019e9164a5c9598f24ecd7796f.pdf.

Büscher, B. (2017). Nature 2.0: Exploring and Theorizing the Links Between New Media and Nature Conservation. *New Media & Society, 18*(5), 726–743.

Christensen, L. T., Morsing, M., & Thyssen, O. (2015). Discursive Closure and Discursive Openings in Sustainability. *Management Communication Quarterly, 29*(1), 135–144.

Clark, T. (1967, January 1). Community or Communities? A Dilemma for Studies of Community Power. *Social Thought and Research*. Retrieved April 1, 2018, from https://kuscholarworks.ku.edu/handle/1808/4662.

Cornelius, N., Todres, M., Janjuha-Jivraj, S., Woods, A., & Wallace, J. (2008). Corporate Social Responsibility and the Social Enterprise. *Journal of Business Ethics, 81*(2), 355–370.

Delmas, M. A., & Burbano, V. C. (2011). The Drivers of Greenwashing. *California Management Review, 54*(1), 64–87.

Floridi, L. (1999). Information Ethics: On the Philosophical Foundation of Computer Ethics. *Ethics and Information Technology, 1*(1), 33–52.

Floridi, L. (2001). Ethics in the Infosphere. *The Philosophers' Magazine*, 6. Retrieved May 1, 2019, from http://blogs.oii.ox.ac.uk/floridi/wp-content/uploads/sites/67/2014/05/eixpm.pdf.

Fox, Z. (2011). 5 Ways That Tech Companies Are Committing to Sustainability. *Mashable*. Retrieved March 1, 2019, from https://mashable.com/2011/11/03/tech-companies-sustainability/#hRwhgsj3Dqqq.

Fujitsu. (2014). *ICT Sustainability: Australian Benchmark*. Retrieved May 10, 2019, from https://www.fujitsu.com/au/about/resources/whitepapers/ict-sustainability-benchmark-au.html.

Graham, M. (2013). The Virtual Dimension. In M. Acuto & W. Steele (Eds.), *Global City Challenges: Debating a Concept, Improving the Practice* (pp. 117–139). London: Palgrave.

Helman, D. (2016). *Berkeley Lab: It Takes 70 Billion Kilowatt Hours a Year to Run the Internet*. Retrieved June 1, 2018, from https://www.forbes.com/sites/christopherhelman/2016/06/28/how-much-electricity-does-it-take-to-run-the-internet/#40b3f2d71fff.

Hickey-Moody, A., & Wood, D. (2008). Virtually Sustainable: Deleuze and Desiring Differentiation in Second Life. *Continuum, 22*(6), 805–816. https://doi.org/10.1080/10304310802452479.

Horst, H. A. (2013). The Infrastructures of Mobile Media: Towards a Future Research Agenda. *Mobile Media & Communication, 1*(1), 147–152. https://doi.org/10.1177/2050157912464490.

Kiron, D., Unruh, G., Kruschwitz, N., Reeves, M., Rubel, H., Meyer, A., & Felde, Z. (2017). *Corporate Sustainability at a Crossroads Progress Toward Our Common Future in Uncertain Times*. Retrieved October 22, 2018, from https://sloanreview.mit.edu/content/uploads/2017/05/d6d6b55daa.pdf.

Market Forces. (2018). *Investing in the Dark*. Retrieved May 9, 2019, from https://www.marketforces.org.au/wp-content/uploads/2018/06/MF-Investing-in-the-Dark-FINAL-A4.pdf.

McManus, P. (2005). *Vortex Cities to Sustainable Cities: Australia's Urban Challenge*. Sydney: UNSW Press.

McLean, J., Lonsdale, A., Hammersley, L., O'Gorman, E., & Miller, F. (2018). Shadow Waters: Making Australian Water Cultures Visible. *Transactions of the Institute of British Geographers, 43*(4), 615–629. https://doi.org/10.1111/tran.12248.

Meadows, D., & The Club of Rome. (1972). *The Limits to Growth: A Report for the Club of Rome's Project on the Predicament of Mankind*. New York: Universe Books.

Merchant, B. (2019). Amazon Is Aggressively Pursuing Big Oil as It Stalls Out on Clean Energy. *Gizmodo*. Retrieved June 10, 2019, from https://www.gizmodo.com.au/2019/04/amazon-is-aggressively-pursuing-big-oil-as-it-stalls-out-on-clean-energy/.

Pasek, A. (2019). Managing Carbon and Data Flows: Fungible Forms of Mediation in the Cloud. *Culture Machine*. http://culturemachine.net/vol-18-the-nature-of-data-centers/.

Plumwood, V. (2008). Shadow Places and the Politics of Dwelling. *Australian Humanities Review, 44*, 139–150. http://australianhumanitiesreview.org/2008/03/01/shadow-places-and-the-politics-of-dwelling/.

Reilly, A. H., & Hynan, K. A. (2014). Corporate Communication, Sustainability, and Social Media: It's Not Easy (Really) Being Green. *Business Horizons, 57*(6), 747–758.

Sadowski, J., & Bendor, R. (2019). Selling Smartness: Corporate Narratives and the Smart City as a Sociotechnical Imaginary. *Science, Technology and Human Values, 44*(3), 540–563. https://doi.org/10.1177/0162243918806061.

Scholz, R. W. (2016). Sustainable Digital Environments: What Major Challenges Is Humankind Facing? *Sustainability, 8*(8), 726.

Shehabi, A., Smith, S. J., Horner, N., Azevedo, I., Brown, R., Koomey, J., et al. (2016). *United States Data Center Energy Usage Report*. Berkeley, CA: Lawrence Berkeley National Laboratory. Retrieved January 15, 2019, from https://eta.lbl.gov/publications/united-states-data-center-energy.

Statt, N. (2018). *Apple Now Says It Is Now Powered 100 Percent Renewable Energy Worldwide*. Retrieved January 14, 2019, from https://www.theverge.com/2018/4/9/17216656/apple-renewable-energy-worldwide-climate-change.

Strengers, Y. (2013). *Smart Energy Technologies in Everyday Life: Smart Utopia?* London: Palgrave Macmillan.
Strubell, E., Ganesh, A., & McCallum, A. (2019). *Energy and Policy Considerations for Deep Learning in NLP* (preprint). Retrieved June 20, 2019, from https://arxiv.org/pdf/1906.02243.pdf.
Wheeler, S. M. (2016). Sustainability Planning as Paradigm Change. *Urban Planning, 1*(3), 55–58.

# 8

# Feeling the Digital Anthropocene

In this chapter, I offer a provocation that talks to current debates on the Anthropocene and the messiness of our digital everyday lives. For the purposes of this chapter, I am opting for the definition of a provocation as an attempt to call forth, rather than to incite anger, although how you respond to this provocation is, clearly, beyond my control. Music also winds its way along this provocation, as the spaces we spend time in are captured by, and played back in, the songs we listen to and make room for. At the same time, music has an agency that is contingent and changes in terms of how it is experienced, and what lines of flight (in a Deleuzian sense) enable it, as Maalsen (2019) has stated. Last, I offer some reflections on the Anthropocene in relation to emotions and affect in digital spaces: a discourse analysis of digital expressions of feeling in the Anthropocene concludes the provocation. The digital is as contingent (Warf 2001) as the Anthropocene (Head 2014), both are networked, material and abstracted spaces. But they are also more-than-real (McLean 2016), working with and producing challenging geographic relations and ways of knowing, and ways of being, that blend rational, emotional and affectual action. Purdy (2015) wrote that he wanted to offer a call to responsibility in *After Nature* and this


J. McLean, *Changing Digital Geographies*,
https://doi.org/10.1007/978-3-030-28307-0_8

provocation joins such an effort—and with a different emphasis. Here, I bring Anthropocene feelings into conversation in this responsibility call, so that different ways of being human and more-than-human are considered at the same time.

The 'digital Anthropocene' is an idea that emphasises the digitization of human–environment relations and changing power relations. Jørgensen writes that 'our standards, for what is natural are distributed and maintained in digital tools and media like databases, computer models, geographical information systems, and so on' (Jørgensen 2014, 109). Digital tools and media are renegotiating our experiences of environments as different opportunities emerge to engage with more-than-humans. Digital Anthropocenes are discussed in this chapter as contingent and multiple: there is no singular way of understanding nature and more-than-humans exert agency in different ways in particular contexts. Travis (2018) suggests that our use of various digital technologies propels people into 'living breathing remote sensors' and not-knowing environmental actors. As this chapter shows, however, there are moments of awareness of the environmental costs of digital lives, as difficult as this may be to see in largely unregulated corporate contexts.

## Troubling the Anthropocene

The impossible Anthropocene that many say we're in—geologists, social scientists and humanities folk—begs for an opening within its global, omniscient framing. In 2000, Paul Crutzen and Eugene Stoermer proposed the Anthropocene concept to capture the scope of human-induced global environmental change (International Geosphere-Biosphere Program 2010). Rather than humans being only subject to global forces of change, including the geologic, it is now proposed that we are also agents of such change. We have moved, the Anthropocene Working Group argues, from the Holocene to the Anthropocene (Carrington 2016). The declaration of a new epoch is required to describe the scale, breadth and depth of the ways we are interfering in multiple systems of which we are simultaneously subject to. The interference is evident in multiple systems, including reduced biodiversity levels, as we see signs of a mass extinction (Barnosky et al. 2011), and in climate change, as we see the effects of a warming climate over all Earth's oceans and land (Intergovernmental Panel on Climate Change 2014).

The Anthropocene idea has been connected to planetary urbanisation; certainly, 'Planetarity is having a moment' (Derickson 2018, 556). Planetarity urbanisation emphasises intertwined economic and political relations and that everywhere is more or less connected to urban systems (Brenner and Schmid 2012). As a feminist scholar, Derickson is concerned that the Anthropocene and planetary urbanisation concepts overly rely on abstractions. Both notions use a species-level, collective form of language: we are all in it together, this global urbanising, we have all produced these conditions, and we are all going to suffer together too. These are sound critiques. But there is also a grounding that comes from gathering evidence of multiple systems collapses to generate scientific and social framings of the Anthropocene.

The Anthropocene idea has begun to enter popular culture, albeit in a modified form in Nick Cave's 'Anthrocene' on his last album, Skeleton Tree. That album was partly written before Cave's son died, too soon, and captures his personal grief entangled with fear for a changing planet. He wrote about how there are powers working in the world that are stronger than humans, and captures the harrowing loss of a child, before connecting this loss to the 'astonishing rise of the Anthrocene': grief, shock, awe together again. This song, embedded in an album full of sadness, links the personal to the global in a way that is echoed by rational scientists not used to sharing such pain in public spaces. It wasn't Cave who made up the Anthrocene, Andrew Revkin first proffered it in 1992 in his book on global warming, but Cave has given it a resonance hard to ignore. Revkin (1992, 55) wrote 'Perhaps earth scientists of the future will name this new post-Holocene era for its causative element—for us. We are entering an age that might someday be referred to as, say, the Anthrocene. After all, it is a geological age of our own making'. The 'our' has been challenged by scholars including Malm and Hornborg (2014) who see the 'Anthropo' as overly generalised and distributing responsibility universally, missing the fact that the wealthiest nations have benefitted from the over-extractive practices that have pushed global environmental systems to the edge.

## Feeling Hopeful in the Anthropocene

There is hope that awareness of the human impact on the earth will translate into human action to mitigate that action; this hope is a significant opening in the Anthropocene that might allow some light in, to paraphrase Leonard Cohen. At the same time, the anthropocentrism that is at the heart of the Anthropocene idea needs to be acknowledged here. Our self-elevation to agents of geological change has a touch of the narcissistic to it, and can be linked to our modernist tendency to continually place humanity outside of, or above, natural processes (Walton and Shaw 2015).

As we might expect, changing environmental conditions do not play out equally for all of us earthly creatures. There are a few animals who are doing well enough in their altered biomes, and introduced species in some parts of the world are thriving. For instance, Bradley Cardinale of the University of Michigan researches how deer are benefitting from cultivated forests and squirrels from parks. These manicured spaces provide a niche habitat for these animals (Baranuik 2017). Cephalopods such as squid are doing relatively well in the Anthropocene, including the Humboldt squid that is growing at faster rates and in booming populations (Doubleday and Connell 2018). These scientific insights capture the 'patchy' aspects of the Anthropocene and may soothe disappointed or deflated souls. But it also shows how keen, and near to desperate, we are to look for hope in these changing times.

Geographer Lesley Head, in her compelling *Grief and Hope in the Anthropocene*, says we need to 're-imagine humans as a force for environmental good, or at least not essentially bad or damaging, as some of the Anthropocene framing suggests' (2017, 11). Stories must be told of humans as being capable of environmental good rather than rampaging recklessly and without thought for the future if a different way of relating to the planet can flourish. The positioning of hope as practice in Head's book offers a constructive way to do this, rather than relying on transitory feelings of optimism to motivate change. Sharing insights into instances of intervention for mitigating the Anthropocene is one way to do this, carrying hope, anger and grief together in a mixed space of diverging emotions.

Beautiful music has been made that captures that mixed emotional terrain that makes up living in, and trying to work against, the Anthropocene. John Luther Adams composed 'Become Ocean' that depicts the changing waters of our world due to melting polar ice and rising sea levels. In an interview following the award of a Pulitzer Prize for his composition, Adams was asked about the tensions that hold his music together:

> '*Interviewer*: Listening to it, I was almost expecting something different. This is music that sort of overwhelms, and you can be absorbed in it, but I don't feel like I'm drowning, you know. I don't feel a sense of alarm — you know, a siren or anything like that.
> *Adams*: We make this distinction between sacred space and quotidian space, but I don't think it needs to be that way. You know, there's this 19th-century idea of the sublime: the idea is that there is an inextricable wholeness to our experience of the world, that contains at once both beauty and terror. And I think I want to be right on that razor's edge.' (NPR Staff 2014)

The obvious problem with living on the razor's edge is the pain that comes with the inevitable cuts; the sublime may hold beauty, yet the terror is real and could encourage humans and more-than-humans to opt for safety. Nevertheless, we can see humans trying to abate the relentless march of the Anthropocene. For example, in Australia we see grassroots initiatives taking up renewable energy as an environmental and economic gesture. Australians have individually taken to solar power in large numbers. According to the Clean Energy Council (2019) small-scale solar produced 19.6% of Australia's renewable energy, and 4.2% of the country's total electricity. Breaking this down in terms of individual households, Bruce and MacGill (2016) state that as of mid-2016, about 16.5% of households have solar panels providing energy to their premises. Australia leads the world with this proportion of households using solar panels. Federal and state government policy, including rebates for solar power installation and tariffs to houses feeding electricity back into the grid via solar power production, has facilitated this.

As Chapter 6 shows, environmental NGOs such as the Climate Council, 100% Renewable and Solar Citizens, have also played a role in helping to facilitate this widespread uptake. A blend, then, of grass-roots-driven action and government policy has begun to produce a more sustainable energy future. One that is disrupting 'traditional' energy supply, to an extent, as we reduce our energy consumption, generally, and increase our renewable energy consumption, specifically.

There is evidence that we can be more green—we have changed our behaviour so that we reduce our environmental impacts in some cases—and we can do so again. For instance, Sydney-siders have reduced water consumption over the past ten years, partly as a result of water-saving education programmes. Rainwater tanks are being taken up in urban spaces as well, another micro-disruption in macro-infrastructure, as households pursue independent, and more economically predictable, water supply for some purposes (Sofoulis 2015). The disruption of big water at the micro-scale demonstrates an expression of agency (Sofoulis 2005) in the Anthropocene that is similar to the uptake of solar panels; people, local councils and community groups are enthusiastic about responding to climate change and water scarcity and are acting in alignment with their interests.

It is crucial to position the shift in relations between human and non-human entities with feelings, to frame such a move where power sits in human and nature relations. Increasingly, rational decisions are not being pursued in environmental and resource practices—the science on carbon emissions has been 'in' for a while now, but places like Australia are still finding it hard to wind back our coal addiction. Clearly, the science is crucial, and it's not all we need. Grappling with these Anthropocene feelings I've been talking about here, ranging from hope to despair, is a crucial step in doing something about the razor's edge, the sublime that Adams referred to in his music on oceans.

Perhaps the way that these changes could continue to emerge include art, such as music, and the redefining of what activism is, and can be. Adams' composition 'Become Ocean' references an 'inviting apocalypse' that may change how humans and non-humans engage. But he disavows activism in writing about his creation:

> My work is not activism. It is art. As an artist, my primary responsibility must be to my art as art—and yet, it's impossible for me to regard my life as a composer as separate from my life as a thinking human being and a citizen of the earth. Our survival as a species depends on a fundamental change of our way of being in the world. If my music can inspire people to listen more deeply to this miraculous world we inhabit, then I will have done what I can as a composer to help us navigate this perilous era of our own creation. (Adams 2015, no page numbers)

Adams is clearly saying he is not making art for activism. Despite this, I think the intent of his art resonates with a shift in activism more broadly, as evident in his statement that human survival depends on transforming our 'way of being in the world'.

To reiterate, there is more than one Anthropocene here (Gibson et al. 2015) and to imagine a smoothly distributed range of impacts arising from global environmental change is naïve. Not everyone is feeling the harsh realities of a too hot Earth in the same way. We know that vulnerability to unwanted impacts of natural disasters is unevenly distributed: those who are already most economically precarious on a global scale tend to be most radically affected by those processes. Further, and as importantly, not everyone is responsible for the production of the Anthropocene in the same way (Cook et al. 2015). If we take climate change, one key aspect of these Anthropocenic times, we know that the most economically wealthy countries are also the ones who have consumed more carbon and pushed our atmosphere into this new state.

Indigenous and allied scholars have argued against the universal framing of the Anthropocene and argued that it is akin to deep colonising. For example, Whyte (2017) sees anthropogenic climate change as an extension of colonisation by settler states. In an overview of the development of Indigenous Climate Change Studies, Whyte argues that decolonising the Anthropocene is necessary. Indigenous people have experienced firsthand disruptive social, political, cultural and environmental forces and:

> Indigenous peoples often imagine climate change futures from their perspectives (a) as societies with deep collective histories of having to be well-organized to adapt environmental change and (b) as societies who must reckon with the disruptions of historic and ongoing practices of colonialism, capitalism, and industrialization. (Whyte 2017, 154)

Holding the Anthropocene at arms-length may be a wise move. Similarly, Donna Haraway has problems with the Anthropocene idea but doesn't throw it away: rather, she offers complementary terms to bolster the paucity of language and concept coming with this notion. As a feminist, cyborg theorist and philosopher of human–animal connections, Haraway advocates the Chthulucene, drawing on science fiction figurations. This alternative nomenclature comes with the slogan 'Make Kin Not Babies!' suggesting that we need to engage with all those around us—humans, non-humans and the inanimate—and connect with these rather than produce more to love. Haraway (2015) states that 'Right now, the earth is full of refugees, human and not, without refuge'. In other words, some of 'us' have rendered strange, at a global scale, that which nurtures us—we could think of the Anthropocene itself as more-than-real in this strangeness.

## Emotion and Affect in the Digital Anthropocene

With the Anthropocene debate firmly in mind, I now gauge how emotion and affect are associated with the Anthropocene in digital spaces. In earlier chapters of this book, we have seen how emotion plays a role in mobilising people and communities to respond to climate change related dilemmas. Here, I look at how the proposition of a new epoch is responded to in the digital, as well as constructed by the digital, to build an understanding of the more-than-real in the Anthropocene. This chapter draws on Twitter as a subset of social media, using the search terms 'I feel' and 'Anthropocene' in a Twitter advanced search, inspired by Pedersen and Lupton (2018) who performed a similar search in a very different context—emotions in mothering on Mumsnet—and gathered useful insights on how digital spaces capture and promote different ways of being. I conducted this search in February 2017 and analysed 100 of the top responses using qualitative content analysis on the posts that were gathered. There are clear limitations in this approach and it is by no means exhaustive. For this provocation, however, it is useful to gain glimpses of what is happening in digital spaces with respect to the Anthropocene. What emerges is an impression of emotion and affect in the digital Anthropocene.

In response to a post on Anthropocene angst from Eric Holthaus (Fig. 8.1), a metereologist based in the United States who has a large Twitter following, Citizen Kelly @flaneurie wrote '"anthropocene angst": what a great and accurate phrase. I feel Anthropocene anxiety, too'. There is black humour mixed in with this expression of angst and anxiety. The visual play on light and dark shadings in the comic are mirrored by the language and wry tone of the original post and respondent.

People on Twitter are contributing to, and drawing directly on, the critiques circulating in the Environmental Humanities that argue for a repositioning of the Anthropocene as a moment of opening, a chance to revisit narratives of environmental destruction and build alternatives (Haraway 2015). For instance, @kateschapira connects emotions about the Anthropocene with those relating to privilege (Fig. 8.2). These terms should be starting points in conversations and actions rather than finishing or closing gestures. After all, the Anthropocene moment can be a generative intellectual event (McGregor and Houston 2018) as well as a process.

The scaling out and in of the emotional terrain from the digital Anthropocene sometimes oscillates between slim hope and outright despair. Figure 8.3 captures @farai discussing her hope that American society can change for the better and move towards greater equality while the Anthropocene looms as an end for our species. Again, the interlinking of social injustices and global environmental change is shown in this impression. @farai draws on humour to round out a profoundly paradoxical relationship to social change and with good reason.

The enormity of the problems emerging with the recognition of this new epoch are understood by those engaging with Anthropocene discourse. Relatedly, issues of moral and ethical complexity are evident in social media dialogues on emotions and the Anthropocene.

**Fig. 8.1** Feeling the Anthropocene—Eric Holthaus (Twitter screenshot, May 2018)

For instance, @Evidentiary1 says that 'I just despair for our planet and its incredible diversity of life that some feel morally superior to #anthropocene #SpeciesExtinction'.

The affect generated by Twitter discourses of the Anthropocene is visible in the likes and retweets, the tweeted responses to declarations of shock and disappointment, fear and concern. We can see in @ EricHolthaus' tweet (Fig. 8.1) that 591 people 'hearted' the tweet and 313 people retweeted it. Twitter's decision in 2015 to afford users with a shift from a star, indicating a favouriting of a tweet, to a heart, produces a relational space with affectual confusion (Bucher and Helmond 2018).

**Fig. 8.2** The generative potential of the Anthropocene (Twitter screenshot, May 2018)

In 2001, Anderson and Smith expressed concern at the silencing of emotions in social research and public life. They argued that emotions are highly political and gendered and yet too frequently ignored. Here, in this Anthropocene moment, we see that emotions are accompanying the science of degradation and damage, affect is circulating in the way we respond to, resist calls of, and re-engage with, global changes that are appearing in our local spaces. The Anthropocene and the digital are both

**Fig. 8.3** Hope and despair in the Anthropocene—Farai Chideya (Twitter screenshot, May 2018)

more-than-real in some ways—challenging ways of knowing and being in the world with ethical and material dilemmas. The digital and the Anthropocene also are generated by ordinary affect, surfacing sometimes invisible processes—such as the code and the global scale—to sit in the space where algorithms and humans meet (Bucher 2017).

The depression and hopelessness that accompanies awareness of the Anthropocene moment need not lead to inaction or passivity. As Osborne (2019) writes, a politics of depression can ground and build communities, as people work together with more-than-humans to imagine future possible worlds, and ways for bringing these into being. The negative may not lead inevitably to destruction if we put grief, anger, frustration, sadness and more, to good use. Using the analogy of gardening and the way we can put sometimes unwanted things to play a role in forging the new, Osborne vividly paints this picture of reworlding: 'I dig ashes, blood, bone, and shit into my garden bed. They help things grow' (Osborne 2019, 147).

The digital Anthropocene also facilitates sharing music to respond to global environmental crises. For instance, Simon Kerr and Christine Parker have co-produced a concert piece called 'Music for a Warming World' (Kerr 2018) that has four movements and is presented digitally (https://www.musicforawarmingworld.org/concert): Storm (science and consequences), Loss (grieving), Change (technology, politics and justice) and Hope (imagination and collective action). Kerr describes himself as a producer, climate thinker and music ian while Parker is a law Professor and VJ who wishes to make an impact beyond academic journals. One piece that forms part of the concert is called 'This Changes Everything' and is inspired by Naomi Klein's (2015) book of the same name. The songs are evocative and arresting; urgency underlies the guitar work and vocals, especially in the 'Change' section.

## Conclusion: Feeling and Doing in the Anthropocene

In Chapter 2 when canvassing the literature relevant to digital geographies in the Anthropocene, I highlighted Gibson-Graham's (2011) argument on how regional development can be supported and facilitated by the vitality and vibrant materiality of the digital. Gibson-Graham asked whether belonging could be rendered differently through new interconnections in digital spaces, with their potential for democratising processes. The vibrant materiality of digital spaces comes partly from their affordances for the simultaneous expression of emotion and reason, where people are able to express shock, anger, hope, disappointment and rage alongside reasoned responses to processes such as changing climates. The more-than-real facilitates the capacity for affect to flow besides, and produce, action of multiple sorts, as Sedgwick (2003) might argue. New ways of representing and understanding data may be useful here, too. For instance, Verhoeven et al.'s (2014) efforts in 'songification' could be applied in communicating the digital Anthropocene, contributing to aural representations of these global environmental processes rather than the familiar visual J curves or planetary systems threshold graphs.

How well we recognise the power of emotion and affect in digital Anthropocene times can possibly transform how this epoch either ends or persists. As Ahmed (2013) shows, emotions have a certain stickiness that can hobble or propel actions. If we ignore the global environmental changes surrounding us, produced by the Global North and experienced most acutely in the Global South, then we perpetuate processes of oppression and marginalisation begun through Imperial and colonial structures, processes and acts. Such denial does not require a reckoning of emotional and affectual terrains in the Anthropocene. But if we take another tack, as so many have done and will continue to do, then complicated, messy emotions come to the fore as decisions to *do something* are prompted. There are global social and environmental justice movements that are asking us to rethink and re-feel the Anthropocene. The possibilities to amplify these in digital spaces are also well known. Nowviskie (2014) summarised these possibilities in her powerful reflection on Digital Humanities in the Anthropocene—there are opportunities afforded by the digital to amplify the voices of human and non-human agents who are working for change.

A friend of mine who is a geologist almost laughed when I first broached the idea of the Anthropocene with her six years ago. I wondered aloud what was funny and she said 'Well, *we* might not be okay, but the Earth, it'll be just fine, it's going to change anyway'. There is a harsh truth in that claim that might urge us to more anger and sadness, or to appreciate the beauty and terror in the sublime as Adams suggests. Yes, this planet will likely outlast humans, and there's the rub, for in doing something for the environment, we are also doing something for humanity as the old dualisms of human/nature just don't work anymore. At least in this formulation, the narcissism that accompanies the Anthropocene framing walks hand-in-hand with efforts to reform human–nature relationships. The more-than-real is a strategy in this context that provides a way to differently understand digital materialities in the Anthropocene.

## References


Adams, J. L. (2015). Making Music in the Anthropocene: How Should Artists Engage in Times of Crisis? *Slate*. Retrieved March 1, 2019, from https://slate.com/culture/2015/02/john-luther-adams-grammy-winner-for-become-ocean-discusses-politics-and-his-composition-process.html.

Ahmed, S. (2013). *The Cultural Politics of Emotions*. Edinburgh: Edinburgh University Press.

Anderson, K., & Smith, S. (2001). Editorial: Emotional Geographies. *Transactions of the Institute of British Geographers, 26*(1), 7–10.

Baranuik, C. (2017). The Animals Thriving in the Anthropocene. *BBC*. Accessed August 1, 2018, from http://www.bbc.com/future/story/20170801-the-animals-thriving-in-the-anthropocene.

Barnosky, A., Matzke, N., Tomiya, S., Wogan, G., Swartz, B., Quental, T., et al. (2011). Has the Earth's Sixth Mass Extinction Already Arrived? *Nature, 471*, 51–57.

Brenner, N., & Schmid, C. (2012). Planetary Urbanization. In M. Gandy (Ed.), *Urban Constellations* (pp. 10–13). Berlin: Jovis.

Bruce, A., & MacGill, I. (2016). *Fact Check: Is Australia the World Leader in Household Solar Power?* UNSW Newsroom. Retrieved December 15, 2017, from https://newsroom.unsw.edu.au/news/science-tech/fact-check-australia-world-leader-household-solar-power.

Bucher, T. (2017). The Algorithmic Imaginary: Exploring the Ordinary Affects of Facebook Algorithms. *Information, Communication & Society, 20*(1), 30–44.

Bucher, T., & Helmond, A. (2018). The Affordances of Social Media Platforms. In J. Burgess, A. Marwick, & T. Poell (Eds.), *The Sage Handbook of Social Media* (pp. 233–253). London: Sage.

Carrington, D. (2016). The Anthropocene Epoch: Scientists Declare Dawn of Human-Influenced Age. *The Guardian*. Retrieved November 10, 2017, from https://www.theguardian.com/environment/2016/aug/29/declare-anthropocene-epoch-experts-urge-geological-congress-human-impact-earth.

Clean Energy Council. (2019). *Clean Energy Council: Solar*. Accessed June 10, 2019, from https://www.cleanenergycouncil.org.au/resources/technologies/solar-energy.

Cook, B., Rickards, L., & Rutherfurd, I. (2015). Geographies of the Anthropocene. *Geographical Research, 53*(3), 231–243.

Derickson, K. (2018). Masters of the Universe. *Environment and Planning D: Society and Space, 36*(3), 556–562. https://doi.org/10.1177/0263775817715724.

Doubleday, Z., & Connell, S. (2018). Weedy Futures: Can We Benefit from the Species That Thrive in the Marine Anthropocene? *Frontiers in Ecology and the Environment, 16*(10), 599.

Gibson, K., Rose, D. B., & Fincher, R. (2015). *Manifesto for Living in the Anthropocene*. New York, USA: Punctum Books.

Gibson-Graham, J. K. (2011). A Feminist Project of Belonging for the Anthropocene. *Gender, Place and Culture, 18*(1), 1–21. https://doi.org/10.1080/0966369X.2011.535295.

Haraway, D. (2015). Anthropocene, Capitalocene, Plantationocene, Chthulucene: Making Kin. *Environmental Humanities, 6*(1), 159–165. Retrieved November 1, 2017, from https://read.dukeupress.edu/environmental-humanities/article/6/1/159/8110/Anthropocene-Capitalocene-Plantationocene.

Head, L. (2014). Contingencies of the Anthropocene: Lessons from the 'Neolithic'. *The Anthropocene Review, 1*(2), 113–125.

Head, L. (2017). *Hope and Grief in the Anthropocene: Re-conceptualising Human–Nature Relations*. London: Routledge.

Intergovernmental Panel on Climate Change. (2014). *Climate Change 2014: Synthesis Report Summary for Policymakers*. Retrieved April 1, 2019, from https://www.ipcc.ch/report/ar5/syr/.

International Geosphere-Biosphere Program. (2010). *Have We Entered the 'Anthropocene'?* Retrieved August 8, 2018, from http://www.igbp.net/news/opinion/opinion/haveweenteredtheanthropocene.5.d8b4c3c12bf3be638a8000578.html.

Jørgensen, F. (2014). The Armchair Traveler's Guide to Digital Environmental Humanities. *Environmental Humanities, 1*, 95–112. https://doi.org/10.1215/22011919-3614944.

Kerr, S. (2018). *Music for a Warming World*. Retrieved from https://www.musicforawarmingworld.org/concert.

Klein, N. (2015). *This Changes Everything: Capitalism vs. the Climate*. New York: Simon and Schuster.

Maalsen, S. (2019). *The Social Life of Sound*. Singapore: Palgrave Macmillan.

Malm, A., & Hornborg, A. (2014). A Geology of Mankind? A Critique of the Anthropocene Narrative. *The Anthropocene Review, 1*(1), 62–69.

McGregor, A., & Houston, D. (2018). Cattle in the Anthropocene: Four Propositions. *Transactions of the Institute of British Geographers, 43*(1), 3–16.

McLean, J. E. (2016). The Contingency of Change in the Anthropocene: More-Than-Real Renegotiation of Power Relations in Climate Change Institutional Transformation in Australia. *Environment and Planning D: Society and Space, 34*(3), 508–527.

Nowviskie, B. (2014). Digital Humanities in the Anthropocene. Retrieved June 10, 2019, from http://nowviskie.org/2014/anthropocene/.

NPR Staff. (2014). *An Inviting Apocalypse: John Luther Adams on 'Become Ocean'*. Deceptive Cadence. Retrieved April 5, 2019, from https://www.npr.org/sections/deceptivecadence/2014/09/28/350911062/an-inviting-apocalypse-john-luther-adams-on-become-ocean.

Osborne, N. (2019). For Still Possible Cities: A Politics of Failure for the Politically Depressed. *Australian Geographer, 50*(2), 145–154.

Pedersen, S., & Lupton, D. (2018). 'What Are You Feeling Right Now?': Communities of Maternal Feeling on Mumsnet. *Emotion, Space and Society, 26*, 57–63.

Purdy, J. (2015). *After Nature*. Harvard: Harvard University Press.

Revkin, A. (1992). *Global Warming: Understanding the Forecast*. American Museum of Natural History, Environmental Defense Fund. New York: Abbeville Press.

Sedgwick, E. K. (2003). *Touching Feeling: Affect, Pedagogy, Performativity*. Durham, NC: Duke University Press.

Sofoulis, Z. (2005). Big Water, Everyday Water: A Sociotechnical Perspective. *Continuum, 19*(4), 445–463.

Sofoulis, Z. (2015). The Trouble with Tanks: Unsettling Dominant Australian Urban Water Management Paradigms. *Local Environment, 20*(5), 529–547.

Travis, C. (2018). The Digital Anthropocene, Deep Mapping, and Environmental Humanities' Big Data. *Resilience: A Journal of the Environmental Humanities, 5*(2), 172–188.

Verhoeven, D. K., Davidson, A., Gionfriddo, A., Verhoeven, J., & Gravestock, P. (2014). Turning Gigabytes into Gigs: "Songification" and Live Music Data. *Academic Quarter, 9*(Autumn), 151–163. Retrieved June 10, 2019, from http://dro.deakin.edu.au/eserv/DU:30068812/verhoeven-turninggigabytes-2014.pdf.

Walton, T., & Shaw, W. (2015). Living with the Anthropocene Blues. *Geoforum, 60,* 1–3.

Warf, B. (2001). Segueways into Cyberspace: Multiple Geographies of the Digital Divide. *Environment and Planning B: Planning and Design, 28*(1), 3–19.

Whyte, K. (2017). Indigenous Climate Change Studies: Indigenizing Futures, Decolonizing the Anthropocene. *English Language Notes, 55*(1), 153–162.

# 9
# Feminist Digital Spaces

Feminist practice and theory is proliferating in multiple spaces, benefitting from the conversations and critical interventions stemming from interactions across social media, activist spaces and academic contexts. In this chapter, I examine the trajectories of social and cultural interventions in more-than-real spaces, and their backlashes, to trace emergent lines of multiply located feminist revitalisations. From #everydayfeminism to #MeToo (Rodino-Colocino 2018), these moments are often growing from social media conversations, reflecting more-than-real possibilities. The digital has afforded spaces for feminist activism that has produced a range of costs and benefits; people and institutions are using digital technologies for reproducing and also challenging sexist and misogynistic behaviours and systems.

The contradictory possibilities of the more-than-real are present at a global scale with the expansion of access to these spaces, and the deepening reach of the digital in everyday life. Feminist digital spaces have arisen as countervailing forces to entrenched sexism and misogyny, in the digital and elsewhere. Diversity is at the core of feminisms today but a common underlying feature of much feminist action is frequent use of, and sometimes reliance on, the digital. The 'call out' culture (Munro 2013) that is facilitated by digital spaces—and specifically social media where users


J. McLean, *Changing Digital Geographies*,
https://doi.org/10.1007/978-3-030-28307-0_9

have some control over their profiles and can curate their feeds—is a key component of this global Fourth Wave. Using 'feminisms' rather than the singular 'feminism' is a deliberate choice in this book to highlight the multiplicity of anti-sexism and anti-misogyny movements. Some read feminisms today as continuing the micropolitics focus of the Third Wave, as this form of feminist action enables individuals to post experiences of sexism and misogyny online and to give voice to their grievances. While digital feminist action has also been framed as 'Fourth Wave Feminism' (for example Clode 2018), I prefer to not use the language of waves to elucidate feminist practices, mostly due to the divisive, categorical and limiting associations of sequential 'waves'. Generational language is associated with this sort of thinking and the commonalities between people of early feminism and more recent feminisms are ignored for the purposes of category maintenance. Sometimes, as in the case of the #MeToo moment that has become a series of movements, feminist activist memes are deployed for particular interventions or issues, where the coalescing of multiple individual stories can contribute to efforts to drive social and cultural change that works across generations. This chapter builds on the long history of feminist theoretical work on emotion and affect (Gorton 2007) and continues the digital work in this area.

## Transforming Feminist Spaces?

Digital technologies are understood as offering possibilities for generative change and recreating, and producing new forms of, oppressions and marginalisations; this book argues that digital geographies contain multiple trajectories, ranging from the oppressive to the liberatory. Of the former, Jons (2013) declares that 'Feminism has left the Academy and spilled into the world wide web' in an article for *Blue Stocking*, an online feminist publication. I think that this enthusiasm is salutary although care needs to be taken to avoid elevating the digital as separate from the digital, and exulting feminist waves potentially renders shifts in feminist practice as disconnected to those that have come before. Clark (2015) analyses the strengths and weaknesses of a range of digital movements to ascertain strategic opportunities for future feminist digital movements. She argues that:

> Digital feminist activism is a new iteration of feminist activism, offering new tools and tactics for feminists to utilize to spread awareness, disseminate information, and mobilize constituents. (Clark 2015, iv)

These affordances of digital feminism are substantive. Digital feminist action is also shaped by corporate entities, such as Twitter and Facebook, that are largely unregulated, raising concerns that this unfettered capitalist context may affect the capacities of those seeking transformative change.

Digital feminist activism frequently emerges from reporting of personal narratives that capture experiences of feminism and sexism. Hashtags have been used on Twitter and Facebook for people to share stories of similar sexual harassment or misogynistic experiences to amass a moment of anger, outrage and sadness—and push for change. For instance, Destroy the Joint is an online feminist group that was formed in 2012 to counter sexism and misogyny in multiple forms and diverse contexts and has its origins in a spontaneous hashtag—#destroythejoint (McLean and Maalsen 2013). It has now got a strong online presence and has been involved in campaigns ranging from stopping debates on rape culture (Keller et al 2018) to 'Counting Dead Women' that is drawing attention to the deaths of women in Australia from gender-based violence. Feminist digital action is pursued globally and uses global digital tools but it is also place-based and partial as a result of uneven access to digital technologies, in part due to digital divides (Fuchs 2009; Van Dijk 2006).

## Digital Feminist Actions in Particular Places

Digital feminisms share similarities with earlier forms of feminist action, focusing on specific forms of sexism in particular contexts while sharing solidarity with a broader movement. For instance, Sills et al. (2016) discuss how social media spaces in New Zealand provide opportunities for pernicious aspects of misogyny to thrive while also facilitating the resistance of these. They interviewed young people on their views of rape culture and what is being done to change aspects of this damaging behaviour. The term 'Rape culture' was introduced in a documentary of the same name in January 1975 produced by Margaret Lazarus (Williams 2015) where critiques of 'phallocentric society' were put forward to show how rape was a part of systemic cultural practices rather than the act of a sole evil perpetrator. Cultural practices including victim-blaming and making jokes about rape are a part of a spectrum of behaviours that form what has been described as rape culture (Horeck 2014; Mendes 2015); the same structures of society that allow sexism to persist also scaffold rape culture.

But as Horeck (2014, 1106) astutely observes 'it is the radical potentialities and limitations of this new temporal regime' that need attention. Horeck (2014) is interested in how hashtag feminism, or digital feminism as it is also commonly known, is refiguring how gender relations are viewed, culturally, socially and politically. Horeck's argument is compelling with respect to the new regime offering new temporal relations, and I add to her claim with the observation that feminist action in the more-than-real is temporal and also *spatial*. Geographers have been adept at drawing attention to the spatial dynamics of the digital (as summarised by Ash et al. 2016) and the spatiality of the digital is an important aspect to these 'new modes' that are sometimes overlooked.

Confronting sexism and misogyny in digital spaces includes challenging rape culture while also providing avenues for its continuation. Sills et al. (2016) state that rape culture is facilitated and challenged by those who engage with the digital. Their interviewees identified a matrix, or 'blanket', of sexism in everyday life where terms like 'rape' are used casually by people to refer to non-sexual harassment practices, and without consideration of the implications of such use on those around them. In Sills et al.'s (2016) project, the researchers and the research participants came from a range of backgrounds and this diversity is reflected in their findings. Māori, Pakeha and Fijian Pakeha researchers with diverse sexualities, from pan to heterosexual, interviewed young people with similarly diverse identities in New Zealand. As well as finding that digital spaces enable oppressive and abusive behaviours to persist, Sills et al. (2016) report how intersectionality is facilitated by the digital and that challenges to sexism are also a part of digital spaces. Social media provides the tools to counter and undermine misogynistic action such as that which comes under the term rape culture, and people are empowered to pursue substantive change because of this technology. In this way, social media is a 'space of civil society' (Fotopoulou 2014, 3). Others argue that digital feminism is not as intersectional as it could be—for example, Clode (2018)—and rightly point to global digital divides as contributing to a lack of intersectionality. While it is important to acknowledge that digital feminism is partial, it does not follow that all digital feminisms are therefore not intersectional.

In Germany, a feminism that builds from and with digital action has gained popularity and provides further support to arguments pertaining to the efficacy of social media in facilitating growing feminist action. It also speaks to the backlash that continues to grow alongside feminist interventions. The hashtag #aufschrei gathered people's experiences of everyday sexism together in one space and was celebrated as the 2013 hashtag of the year in the German-speaking world; aufschrei translates to 'yell' or 'outcry' in English. Drüeke and Zobl (2016) describe how discourse circulated with the hashtag #aufschrei included recognition of the multiplicity of feminist action, in terms of campaign focus, modes of engagement and participants' skills and identities. They quote one participant on why #aufschrei was so successful:

> A political action is successful when it draws the *attention*, *resources*, and *time* of as many people as possible, and that works only when debate occurs at several different levels and from a multitude of perspectives as this allows everyone to find the appropriate place to join in. (antjeschrupp, blog entry from February 2, 2013, quoted by Drüeke and Zobl 2016, 49)

The call-out quality of #aufscrhei is a form of digital activism common to digital feminisms, in allowing individual, sometimes small-scale acts of sexism to be expressed and supported collectively. Worth highlighting from antjeschrupp's observation is the spatial element of effective social movements: the diversity of digital feminism provides people a 'place to join in'. The more-than-real allows for that place to emerge as well as for it to be challenged. As with any vocalisation of injustice, #aufschrei attracted dissent: Drüeke and Zobl (2016) report how within two weeks of the hashtag emerging, twenty percent of tweets were undermining or directly attacking feminist thought and practice.

Backlash is, of course, nothing new in feminist discourse (Faludi 1991) but the different opportunities that the digital allows make people vulnerable in new and alarming ways. The narratives that people share to join together under hashtags or more formal movements can be traced back to individuals and institutions and, in the process, these people and/or groups can be exposed to harm. In this way, one of the major strengths of the more-than-real is also a significant weakness.

Digital feminist action in China is connecting to, and in turn remaking, the diffuse #MeToo movement. Luo Xixi was the first Chinese woman to publically and digitally refer to the #MeToo movement as an inspiration for her to share her experience of sexual harassment in Beihang University in Beijing. Luo circulated her story on multiple platforms, including Weibo—China's version of Twitter—and has spoken of how global conversations about sexism were instrumental to motivating her to seek justice (Qiqing 2018). There are signs that more people are drawing inspiration from the more-than-real dialogue on sexism and sexual harassment: Chen and Zhang (2018) describe how digital technologies were used in July 2018 to accuse a dozen men of different forms of sexual harassment. In lieu of law reforms to support victims of sexual harassment, the more-than-real is providing avenues for seeking justice. Ho and Tsoi (2018, paragraph 7) observe that 'Chinese feminist groups are encouraging women to follow in Ms Luo's footsteps, but think that even if the #MeToo movement takes off it will look quite different in China'. The importance of local and national specificities is visible in the iterations of #MeToo emerging in different social and cultural contexts. Chinese government censorship extends into social media spaces and Zeng (2018) writes how the #MeToo movement has resisted such censorship by using alternate language to identify feminist digital activism. In 2018, Zeng (2018, no page numbers) notes that 'the primary hashtag of China's #MeToo campaign – #MeTooInChina – was temporarily blocked. In response to this, Weibo users launched the alternative hashtag #RiceBunnyInChina to continue the campaign'. 'Rice Bunny' is pronounced as 'mi tu' in China (Zeng 2018) and has become a hidden way to bring together women who have experienced sexual harassment or assault. Playing with foreign languages in digital spaces has enabled Chinese feminists to continue their anti-sexism campaigning.

Like other digital activism, digital feminist action in Mexico involves protests against the violence that women and children experience, including under the #NiUnaMas/NiUnaMenos (not a single woman more/less) slogan (Lopez 2018). Despite these efforts, the murders of women continue, as does the lack of punishment of those perpetuating these crimes. The most intense of these injustices is concentrated in Ciudad Juárez where femicide occurs (López 2018; Wright 2011). Femicide is an extreme expression of patriarchal force: it is the killing of women by men because of their gender (Musgrave 2016). López (2018, 3) writes that this femicide pandemic 'must be placed in a national context of uncontrolled violence from organised crime, impunity, institutional corruption, and a patriarchal mentality'. Geographer Melissa Wright researches the necropolitics of femicide and drug-related deaths in the Juárez region, arguing that a particular politics of gender is at work here that should be given more attention (Wright 2011). Necropolitics is intertwined with biopolitics, according to Wright (2011) and Mdembe (2003), as the work of death (necropolitics) is co-produced in the work of life (biopolitics). Analysis of necropolitics and biopolitics in Juárez has not included sufficient consideration of gender relations, according to Wright (2011). While there are offline protests against this violence, it does not seem that gender relation issues are as strongly attended to in these protests. Networked publics (boyd 2010) may support the offline protests to cross thresholds of inattention towards gendered violence but fundamental support from the state is clearly necessary too.

Digital feminist action in different nations of Africa is quite fractured, partly due to the digital divide and histories of colonisation (Tagny 2016). In west and central Africa, feminist activists have struggled to secure affordable access to digital spaces and cross language barriers to build regional allegiances on feminist issues. However, diffuse efforts to address sexist and misogynistic behaviour is evident as Salami (2014) notes in her account of how African feminists are harnessing the possibilities of digital spaces. As a Nigerian-Finnish writer, Salami argues that there are examples of people using social media and blog spaces to share narratives of independence, challenge child marriage and achieve political power. For instance, Yasmina Oue'gnin is a female Member of Parliament in Cote d'Ivoire who used social media, including Facebook and Twitter, to support her campaign for election. Salami (2014, paragraph 9) believes that 'Oue'gnin managed to charm and engage people and ultimately became one of the youngest members of parliament in the country's history, as well as a champion for online citizen engagement in democratic processes'. In an earlier commentary piece for the *Guardian*, Salami (2013) notes the diversity of feminist action in Africa and the success of strong women in political representation in countries such as Rwanda where the 64% of government representatives are women.

In South Africa, social media has facilitated campaigns to counter rape culture, for example with #EndRapeCulture (Gouws 2018). This campaign involved topless marches by women students in university grounds and cities of South Africa to voice their anger at experiences of rape and the police's habit of ignoring these crimes. Three quarters of women in South Africa experience some form of violence and abuse (Dranzoa 2018). Gouws (2018, 3) writes that

> When the campaign started out most of the organisers were black African women, taking protests to the streets, often bare breasted or in underwear to say that "enough is enough – we are tired of this violence!"

Furthermore, these protesters embraced being 'feminist' as an identity—something that earlier South African generations of women had eschewed according to Gouws (2018). These feminists are actively challenging colonial discourses that frame South African sexuality, similarly challenging colonial intimacies as Indigenous women have in Australia (Sullivan 2018).

Sexual harassment and assault are experienced by many people but rarely reported to the police, while prosecutions of perpetrators are also infrequent. In Australia, national statistics of sexual assault rates and reports to police show that 9 out of 10 women who were victims of sexual assault did not contact the police about their experiences (Australian Bureau of Statistics 2016). Bhattacharyya (2018) notes that experiences of sexual harassment in the workplace are underreported and that as many as 70% of women do not report such incidents. Bollywood has yet to experience its #MeToo movement, according to Bhattacharyya (2018) and feminist digital action in India is emergent. Sujatha Subramaniam's thesis (2014) 'From the Streets to the Web: Feminist Activism on Social Media' examines the entanglements between feminist activism and digital technologies, considering whether feminism can change the masculinism associated with digital technologies, and vice versa, in India. Subramaniam (2014) interviewed people involved in digital activism from different classes and geographic areas in India and offers a thoughtful analysis of what activism is and how it is framed by people pursuing it. One research participant pointed out the inconclusive aspects of defining digital activism:

> Shobha S.V., who works at a Delhi-based NGO, reflects on whether she sees herself as an activist and says, "My interaction with social media came primarily out of a need to express myself, like an online diary. But while all issues were personal, I also wanted the space to reflect mostly my political views. I suppose that does make me an activist. The hesitation is not because I don't see myself as an activist but I am trying to figure out what activism is." (Subramaniam 2014, 3)

Reflexivity is core to digital feminist action, or activism, and critical thinking and self-expression runs through many of these examples of digital efforts to pursue change. There is time and space for people to explore the meaning of their digital actions in issue publics (Kim 2012). At the same time, digital democracy and reliable access to digital spaces are overarching issues that challenge efforts for feminist activists to expand their digital work in India. Subramaniam interviews activists including Archismita Choudhury, a student and moderator of a feminist Facebook page, who highlights the potential for social media to support the capacity for people to have their voices heard: 'Protests can only take you so far in terms of visibility. With digital technology, you have networks from around the world participating' (Subramaniam 2014, 5). Digital feminist action is clearly intertwined with traditional forms of protest—including protests on the street (Gerbaudo 2012). Visibility and, implicitly, power, is increased through drawing on the potentials of the more-than-real.

## Transforming Gender Relations, Transforming the Digital?

Challenges to the patriarchy are being made across the world with converging patterns: calling out sexism and misogyny in digital spaces, drawing on personal stories, building large-scale mainstream social movements. The similarities in the collection of feminist digital actions show how it is more than 'ad hoc issue publics' (Bruns and Hallvard 2014) that are being formed on social media to counter sexism. Common to these differently positioned social movements already mentioned in this chapter built from the more-than-real, from India to Mexico, is the tendency to generalise from specific experiences and share narratives that can make a difference. As Drüeke and Zobl (2016) point out, the coming together of multiple stories under one hashtag increases the visibility of individual experiences and has the potential to draw unwanted negative attention to personal experiences. At the same time, however, such attention is crucial for the critical construction of a vibrant social movement. The downfalls of participating in a broader movement in the paradoxical social and cultural spaces found in the more-than-real can not be glossed over here. And nor should the benefits.

Participation in digital feminist action may be linked to gendered use of social media. Munro (2013) argues that there is evidence of women leading the way in participating in the digital: in Turkey, for example, 72% of social media users are women (Munro 2013, 23). If feminist digital action is arising from social media, and more women are using social media in particular circumstances, then it makes sense that these diverse efforts to end sexism are emerging. Eslen-Ziya (2013) describes how Turkish feminism has developed through the use of networked publics as activists resisted Turkish President Erdogan's condemnation of abortion. Activists used an online petition to protest the proposed reduction of time for legal terminations from ten to four weeks' pregnancy, making the window to secure a wanted termination near impossible. Turkish feminists started an online campaign entitled 'My Body, My Choice' and activists reached out to collaborators around the world: 'On 6 June, 2012 the European Women's Lobby (EWL) an umbrella organization of women's associations in the EU, posted the following on its Facebook page: Say No to Abortion Ban! Support women in Turkey!' (Eslen-Ziya 2013, 867). Facebook groups formed, including the 'Kürtaj yasasına hayır insiyatifi' (Initiative against the law to ban abortion) group and, combined with other political strategies, these efforts were ultimately successful as the Turkish government dropped the anti-abortion legislation. Turkish feminism has incorporated efforts to challenge patriarchal power structures from other contexts to support national causes. There is a multi-scalar element to digital feminist action that supports and deepens the efforts of local and national campaigns.

## Organising More-Than-Real Feminisms with Emotion and Affect

Feminist action in the more-than-real is multiple: at times and in certain spaces, it can involve careful curation, direction and engagement—often this work is without payment or recognition—while it can also be ad hoc, unorganised and fleeting. The people who do the curation work of managing Facebook feminist groups or web sites to support feminist action, perform labour that is often invisible and that extends the reach of the digital. Kylie Jarrett (2017) describes the unpaid and unrecognised work of digital media as akin to a 'Digital Housewife', drawing on critiques of unpaid domestic work. Commercial digital entities rely on this unpaid labour to (re)produce their everyday relevance. The Digital Housewife is not meant to be a gendered term, despite its clear reference to women's work, as Jarrett (2017) uses the term to offer a Marxist critique of the way people of all genders engage with the digital who increase the economic value of the digital by generating and sharing affective experiences, while not being remunerated.

The online movements involved in feminist action rely on digital labour, facilitating the networking of people and groups in attempts to force change. Research has shown that emotional motivations propel such work, including by Gleeson (2016) who discusses the emotional labour involved in facilitating feminist ethical and effective digital spaces. Drawing on interviews with leaders of digital feminist spaces, Gleeson reports that there is enjoyment and frustration mixed into the job of moderating online harassment, accompanied by substantial amounts of burnout and stress. The emotional work of processing inappropriate and offensive posts pushes many moderators to burnout: emotional digital labour, even when performed in shared shifts, can wear people down. The forms of this emotional labour for moderators included positive and negative aspects according to Gleeson (2016, 82):

> a number identified that they did it for the "love" of the job (Irene), that they were "passionate" about the cause (Pam), and "really offended" (Hannah) or "pissed off" by media representations of women (Gina)

Gendered divisions of labour in terms of reproductive and paid work are well understood (Sanchez and Thomson 1997). With respect to the more-than-real, Arcy (2016) outlines an important feminist project to examine the way that gendered divisions of labour is extending into digital spaces. Women perform gendered labour in face-to-face work contexts that allow people to feel comfortable and produce seamless workplaces. This labour, Arcy argues, is extended in the digital where 'women invest their emotional energy to engage with and generate content that in turn adds value to branded platforms' (Arcy 2016, 2). Labelled 'immaterial digital labour', Arcy suggests that simple acts such as liking a Facebook post relies on accumulation and repetition. Affect is perceptible in the generation of multiple likes and the impact of expressing emotions in these seemingly simple ways is inbuilt to the capitalist success of digital corporations. The digital immaterial labour is framed in this context as a burdensome load where ugly feelings can potentially overwhelm (Ngai 2005) but, as Gleeson shows, there are also *generative* possibilities of the affective labour in building and maintaining digital spaces.

The micro-scale experiences of sexism are a general focus of digital feminist action as well and interventions exist that show the cumulative impact of mundane instances of everyday sexism. For example, in 2012, Laura Bates launched the Everyday Sexism Project

(Bates 2015) and invited people to share their narratives of harassment on her website and with the hashtag #everydaysexism on social media (see Fig. 9.1). Bates (2014) wrote a book about the development of the website and recounts how within 18 months of its launch, it had reached 18 nations around the world. The inspiration for creating the site came from a multitude of micro-scale and *ordinary* experiences of sexism that Bates had experienced and heard about others experiencing: everyday moments of sexual assault and harassment that had become normalised. Bates no longer wanted to accept that these were, in fact, acceptable incidents and that by sharing and documenting stories of everyday sexism, the visibility of these banal and intrusive acts might

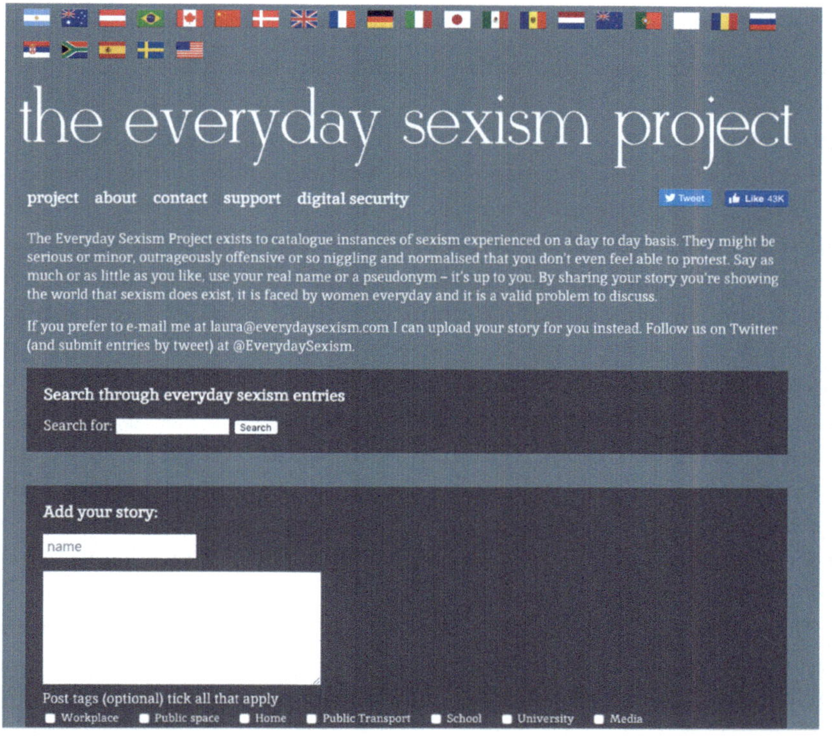


**Fig. 9.1** The interface of everyday sexism (https://everydaysexism.com/)

shift their social and cultural framing from meaningless to intolerable. Bates purposely kept the definition of everyday sexism open to allow stories of gender imbalance to sit alongside more harrowing stories of assault and rape to extend the boundaries of the project and allow the continuum of sexism's manifestation to be expressed. The aim was to not diminish the impact of minor incidents of sexual harassment or gendered discrimination but rather to counter how frequently these are cumulatively ignored or overlooked. The ordinariness of micro-scale sexisms surge together in Everyday Sexism to produce affect that is hard to ignore, 'little somethings' that can move towards transformations, 'dreamed up in the course of things' (Stewart 2007, 9).

## The Motivations of Digital Feminist Activists

Given the costs associated with engaging with more-than-real spaces, it is worth considering the benefits that accompany efforts to pursue feminisms therein. Psychological research has shown that 'using Twitter as a collective action may enhance women's well-being after being exposed to sexism' (Foster 2015, 640). Foster's research involved tracking how people responded to digital narratives on Twitter and recorded their affectual responses following exposure to sexist material. Those who shared public tweets about their experiences reported reduced negative affect and felt that they had an improved sense of well-being. So according to this psychological research, the decision to take decisive digital action makes at least some feminists *feel better*, regardless of the outcomes of such work. Further, in Foster's study, tweeting about sexism tended to speak to a collective audience; the critiques of sexism were outward-looking and intended to generate solidarity and identification from and with others. The support of others improved the way that those subject to such acts felt about them and served to provide a type of informal justice.

Street-based sexual harassment is rarely addressed through criminal justice systems and so Fileborn (2014) conducted empirical analysis of feminists' informal justice-seeking behaviours in digital spaces. The aforementioned Everyday Sexism Project and Hollaback! have provided opportunities for sharing narratives of harassment and, especially in the case of Hollaback!, for people to collectively counter these behaviours. Both sites begin with the principle that facilitating people to give voice to their experiences can be useful and generative. For instance, Everyday Sexism's Laura Bates argues that expressing narratives is important:

> Don't underestimate the catharsis and empowerment that can come simply from telling your story and having it accepted and believed, in a world where it is so often ignored or brushed off. (Bates 2015, paragraph 8)

Digital storytelling is one of the powerful tools that feminists who are agitating for change employ. The possibility of expressing anger, disbelief, and sadness, with an expectation that one's story will be received in a genuine way seems to underpin the engagement with such spaces. Similarly, social media sites provide opportunities for storytelling that have multiple effects—and some may be positive or negative. As de Jong (2015) found in her study of digital storytelling and musical and queer festivals in Australia, there are strengths in using social media for sharing narratives of self-expression. Her research involved working from within Facebook and friending her research participants, forming research relationships over a longer period of time, and thereby allowing participants to take on powerful roles in shaping the research process and becoming co-producers of the work. In other contexts, such as feminists working in Turkey to maintain access to safe and reliable contraception, the stakes are different and a focus on governance interventions crucial, making co-production of research far from a priority.

Answering the why of digital feminist activism brings complex and nuanced explanations, especially when the backlash to such action is often strong. Nakamura (2015) argues that women of colour provide their free labour to improve digital spaces but the response from others to such interventions is frequently aggressive and resistant to change, accusing those 'calling out' such behaviours of policing free speech and pursuing politically correct agendas. The more-than-real allows for people who are unable to obtain justice elsewhere to seek some alternative avenues. Clark (2018) states that the '#MeToo movement is a symptom of the fundamental failures of our criminal justice system to protect the rights of victims'. The unsuccessful efforts of people to do something about sexual harassment and assault, and the frustrations with failure of criminal courts to secure prosecutions, has prompted much of the social media-driven interventions. Traditional media are also joining these efforts, as demonstrated by the 4 Corners show 'I am that Girl' in May 2018 which shared Saxon Mullins' story of sexual assault and the unsuccessful criminal prosecution that ensued (4 Corners 2018). The remarkable documentary was received with widespread anger at the way the formal criminal justice system treats victims of assault and ultimately prompted a review by the New South Wales Attorney General into consent laws (Mulligan and Carter 2018). Mullins' feminism sustained her through a traumatic process and galvanised her to seek further action besides that which the New South Wales court system could provide.

## The Possibilities and Constraints of Digital Feminist Action: Besides #MeToo?

As this chapter has outlined, digital feminist action has been emerging in different parts of the world, along with the spread of digital technologies. Sometimes digital feminist moments and movements are endogenous to digital spaces while in other instances digital spaces have facilitated the growth of networked feminist publics, to work with boyd's (2010) useful influential term. Since 2017, a global interest in the politics of call-out culture and the widespread experiences of sexual harassment has gathered under the #MeToo umbrella, yet it began well before then; Tarana Burke started a not-for-profit organisation to support African American women who had experienced sexual harassment and assault called 'Me Too' in 2006 (Regulska 2018; Mendes et al. 2018). When Alyssa Milano tweeted her original 'me too' post, she did not reference the pre-existing organisation but quickly did when it was drawn to her attention. Milano's tweet in October 2017 was 'If you've been sexually harassed or assaulted write "me too" as a reply to this tweet'. The hashtag was created as the replies to her tweet grew. Tarana Burke and Alyssa Milano have since spoken together about their shared interest in addressing sexual harassment and assault (Chan 2017). Over 1.7 million people have shared their stories of sexual harassment and assault under the hashtag #MeToo and one of the perpetrators that prompted the sharing of these narratives, Harvey Weinstein, has experienced career setbacks and faced criminal charges for sexual assault (Jaffe 2018). Digital feminisms are diverse, spatially and temporally, and focus on different issues but all tend to draw on digital and non-digital actions to look for change. Further, digital feminists tend to share in interest in committing to this diversity: there is an openness and broadness that underpins these efforts.

The #MeToo moment/movement shares commonalities with other feminist digital movements including with respect to sharing an interest in multiple forms of social and cultural change. Pitman (2018) argues that we can understand misogyny as a 'human pyramid' rather than a spectrum. A pyramid is formed when 'minor acts [of sexual harassment or assault] support the major by providing, at best, a foundation of blithe indifference, and at worst an atmosphere of amusement at the denigration of women' (Pitman 2018, no page numbers). Pitman extends this metaphor from the base, where silence is offered in response to assault, to another level where men join in on the joke, to the peak where men behave with impunity towards women, such as Weinstein and Australia's Don Burke. Pitman suggests that the pyramid works like a food chain, where each level is crucial to supporting the one above and below. The more heinous offences of gender-based violence are supported by casual mistreatment and nodding humour at misogynistic statements.

The 'e-bile' that women experience disproportionately to men is a part of the lower, broader base of the pyramid of misogyny. Rather than ignore the gendered vitriol that women experience, sociologist Jane (2014) says that we must turn towards and confront the users of such language and violent imagery to change the behaviours. Her powerful article uses direct quotes of abuse from social media, such as in her title 'Back to the kitchen, cunt' and 'Wash the encrusted cum stanes (sic) off your geriatric fathers beard', to highlight the vindictive and appalling quality of the discourse that circulates online. The purpose of Jane's research is to shine an (academic) light on the everyday gendered abuses happening in digital spaces, mirroring the work of many feminist activists online, as its tenor is increasing and its commonplaceness concerning. Relating to the latter, '[o]ne of the perverse paradoxes of e-bile is that the most personal of insults, attacks and threats can seem generic, predictable and almost tedious as a result of their ubiquity' (Jane 2014, 566). Doyle (2011) collated instances of digital abuse that were shared under the #MenCallMeThings and it presents a depressing catalogue of the gendered vitriol that women experience in the digital. However, she argues that by drawing these together and categorising the different types of aggression directed towards women, the strength of the collective is highlighted because no matter what is said to women in the more-than-real, the ease of fighting back in the digital means that shaming and calling out behaviours far exceed the initial abuse. The paradoxical qualities of digital spaces are found within e-bile here in the bitter and destructive end of digital behaviours. I read these as the accreting of everyday affect to produce negative excesses: just as the digital facilitates positive more-than-real experiences, so too can it render negative landscapes of emotion and affect.

The confession culture that accompanies much digital feminist action has its personal and social costs as well. Those subject to sexism and misogyny carry the burden of speaking up against it but connections and alliances to counter these practices are crucial to overcoming this cost. As Ahmed (2015, 9) argues 'Connections can be what we have to struggle for, because there is so much silence about sexism: sexism makes it costly for women to speak about sexism'. The double bind of having to call-out sexist acts and institutions and then being met with either silence or disbelief due to the already-existing sexism compounds and concretises sexism.

In another perspective on the costs of the more-than-real, Nakamura (2015, 106) unequivocally states that 'cheap female labour is the engine that powers the Internet'. As a result, the extractive corporations that profit from this lowly paid (and often unpaid) work are the biggest beneficiaries of the more-than-real. Further, it is women of colour who provide the majority of this labour and Nakamura (2015) argues that digital labour has to be inexpensive in order for digital culture to even exist. This labour travels along the commodity chain—people are involved in reworking these spaces at every stage of the process of making the digital, from production of the devices that facilitate digital culture to the moderating and calling-out of racism, sexism and ableism in the digital. Nakamura (2015) calls these labourers venture community managers as their interventions aim to make better conditions for all in gaming and social media worlds—similar to early content producers and managers.

## Conclusion: Affective Contagions

From China to India, South Africa to Turkey, digital feminist action is happening. The more-than-real manifests in different ways depending on a number of factors including particular governance contexts, context-specific legal regimes, experiences of sexism and opportunities to use digital technologies. There is no one 'Fourth Wave' happening now that stems from digital feminist action; it is too diffuse, multiple, diverse and contingent. Rather, there are radically different feminist actions taking advantage of the affordances provided by digital spaces, sharing some features with each other and simultaneously geographically, socially, culturally and politically distinct. The more-than-real facilitates this contradiction—where global connections might provide moments of solidarity but these are sharply renegotiated by the contexts and priorities of particular feminist interventions.

The processes pursued by digital feminist activism are sometimes organic and at other times carefully strategised. The spontaneous drive behind hashtags like #everydaysexism and #MeToo could be explained by affective contagions (Gibbs 2001). In writing about the Australian right-wing politician Pauline Hanson and her supporters, Gibbs states that 'Bodies can catch feelings as easily as catch fire: affect leaps from one body to another, evoking tenderness, inciting shame, igniting rage, exciting fear – in short, communicable affect can inflame nerves and muscles in a conflagration of every conceivable kind of passion' (2001, paragraph 1). In everyday, face-to-face interactions, such moving of affect is understandable. Being with people and sharing embodied emotions can lead to a transmission and elevation of affect.

How does this translate in digital spaces where interactions are mediated by technologies and platforms that transform, distort, reflect and amplify? Gibbs argues that media intensify and elevate affectual processes by providing avenues to circulate information and narratives:

> Not only do they act as amplifiers of affect, heightening and intensifying affects (by amplifying the tone, timbre and pitch of voices and, in the case of television, by means of close ups which provide a concentrated focus on facial expressions), but of course they dramatically increase the rapidity of communication of affect, and they extend its reach to the point where it is now almost global. (Gibbs 2001, paragraph 1)

While Gibbs was talking about pre-social media devices, similar processes are clear for digital spaces. By amplifying the tone, timbre and pitch of voices in digital spaces, and reducing spacetime differences on a global scale, digital spaces intensify affects and give forward momentum to spontaneous interventions.

The tools that facilitate digital feminist action are nominally global but they are not necessarily universally accessible, and so limits and differences exist in terms of how well people are able to engage with, and commit to, digital activism for feminist ends. Relatively early calls for the Information and Computer Technology (ICT) sector to peel back the myth of technology neutrality (e.g. Marcelle 2006), which argued that technologies are universally available and useful, seem to have not further developed into sufficient changes in this area. Returning to Sedgwick's thinking on nondualism, rather than reproducing binaries in this overview of digital feminisms from around the globe, the more-than-real sits with ideas about how digital geographies are made, and are making us, in sometimes surprising ways.

## References

Ahmed, S. (2015). Introduction: Sexism—A Problem with a Name. *New Formations: A Journal of Culture/Theory/Politics, 86*(1), 5–13.

Ash, J., Kitchin, R., & Leszczynski, A. (2016). Digital Turn, Digital Geographies? *Progress in Human Geography, 42*(1), 25–43. https://doi.org/10.1177/0309132516664800.

Australian Bureau of Statistics. (2016). *4906.0—Personal Safety, Australia, 2016*. Retrieved August 1, 2018, from http://www.abs.gov.au/ausstats/abs@.nsf/mf/4906.0.

Arcy, J. (2016). Emotion Work: Considering Gender in Digital Labor. *Feminist Media Studies, 16*(2). https://doi.org/10.1080/14680777.2016.1138609.

Bates, L. (2014). *Everyday Sexism*. London: Simon and Schuster.

Bates, L. (2015). Sharing Stories of Sexism on Social Media is 21st Century Activism. *The Guardian*. Retrieved August 2, 2018, from https://www.theguardian.com/lifeandstyle/womens-blog/2015/feb/06/sharing-stories-of-sexism-on-social-media-is-21st-century-activism.

Bhattacharyya, R. (2018, March). #MeToo Movement: An Awareness Campaign. *International Journal of Innovation, Creativity and Change, 3*(4), 1–12.

boyd, d. (2010). Social Network Sites as Networked Publics: Affordances, Dynamics, and Implications. In Z. Papacharissi (Ed.), *Networked Self: Identity, Community, and Culture on Social Network Sites* (pp. 39–58). London: Routledge.

Bruns, A., & Hallvard, M. (2014). Structural Layers of Communication on Twitter. In K. Weller, A. Bruns, J. Burgess, M. Merja, & C. Puschmann (Eds.), *Twitter and Society* (pp. 15–28). New York: Peter Lang.

Chan, M. (2017). 'Now the Work Really Begins.' Alyssa Milano and Tarana Burke on What's Next for the #MeToo Movement Time. *Time*. Retrieved June 5, 2018, from http://time.com/5051822/time-person-year-alyssa-milano-tarana-burke/.

Chen, T. P., & Zhang, C. (2018). #MeToo Movement Gathers Momentum in China. *Wall Street Journal*. Retrieved April 1, 2019, from https://www.wsj.com/articles/metoo-movement-gathers-momentum-in-china-1532793600.

Clark, C. (2015). *#TrendingFeminism: The Impact of Digital Feminist Activism*. Masters Research thesis, the Faculty of the Columbian College of Arts and Sciences of the George Washington University. Retrieved May 1, 2019, from https://pqdtopen.proquest.com/doc/1691802140.html?FMT=ABS.

Clark, K. (2018). #MeToo Exposes Legal Failures but Trial-by-Twitter Isn't One of Them. *The Conversation*. Retrieved July 5, 2018, from http://theconversation.com/metoo-exposes-legal-failures-but-trial-by-twitter-isnt-one-of-them-92683.

Clode, L. (2018). *Post 9 of #GPC@25: A Defence of Fourth Wave Feminism? By Lucy Clode*. Retrieved April 4, 2019, from https://genderplaceandculture.wordpress.com/2018/08/22/post-9-of-gpc25-a-defence-of-fourth-wave-feminism-by-lucy-clode/.

De Jong, A. (2015). Using Facebook as a Space for Storytelling in Geographical Research. *Geographical Research, 53*(2), 211–223.

Doyle, S. (2011, November 10). But How Do You Know It's Sexist? The #MenCallMeThings Round-Up. *Tiger Beatdown*. Retrieved November 7, 2018, from http://www.tigerbeatdown.com/2011/11/10/but-how-do-you-know-its-sexist-the-mencallmethings-round-up/.

Dranzoa, C. (2018). Sexual Harassment at African Higher Education Institutions. *International Higher Education, 94,* 4–5. https://doi.org/10.6017/ihe.2018.94.10513.

Drüeke, D., & Zobl, E. (2016). Online Feminist Protest Against Sexism: The German-Language Hashtag #aufschrei. *Feminist Media Studies, 16*(1), 35–54. https://doi.org/10.1080/14680777.2015.1093071.

Eslen-Ziya, H. (2013). Social Media and Turkish Feminism: New Resources for Social Activism. *Feminist Media Studies, 13*(5), 860–870. https://doi.org/10.1080/14680777.2013.838369.

Faludi, S. (1991). *Backlash: The Undeclared War Against Women*. Portland, OR: Crown.

Fileborn, B. (2014). Online Activism and Street Harassment: Digital Justice or Shouting into the Ether? *Griffith Journal of Law & Human Dignity, 2*(1), 32–53.

Foster, M. D. (2015). Tweeting About Sexism: The Well-Being Benefits of a Social Media Collective Action. *British Journal of Social Psychology, 54*(4), 629–647.

Fotopoulou, A. (2014). Digital and Networked by Default? Women's Organizations and the Social Imaginary of Networked Feminism. *New Media Society, 186,* 989–1005. https://doi.org/10.1177/1461444814552264.

Fuchs, C. (2009). The Role of Income Inequality in a Multivariate Cross-National Analysis of the Digital Divide. *Social Science Computer Review, 27*(1), 41–58.

Gerbaudo, P. (2012). *Tweets and the Streets: Social Media and Contemporary Activism*. London: Pluto Press.

Gibbs, A. (2001). Contagious Feelings: Pauline Hanson and the Epidemiology of Affect. *Australian Humanities Review*. Retrieved August 5, 2018, from http://australianhumanitiesreview.org/2001/12/01/contagious-feelings-pauline-hanson-and-the-epidemiology-of-affect/.

Gleeson, J. (2016). '(Not) Working 9-5': The Consequences of Contemporary Australian-Based Online Feminist Campaigns as Digital Labour. *Media International Australia Incorporating Culture and Policy, 161*(1), 77–85.

Gorton, K. (2007). Theorizing Emotion and Affect: Feminist Engagements. *Feminist Theory, 8*(3), 333–348. https://doi.org/10.1177/1464700107082369.

Gouws, A. (2018). #EndRapeCulture Campaign in South Africa: Resisting Sexual Violence Through Protest and the Politics of Experience. *Politikon, 45*(1), 3–15. https://doi.org/10.1080/02589346.2018.1418201.

Ho, G., & Tsoi, G. (2018). Will #MeToo Spread in China? *BBC News*. Retrieved June 10, 2019, from https://www.bbc.com/news/world-asia-china-42577654.

Horeck, T. (2014). #AskThicke: "Blurred Lines", Rape Culture, and the Feminist Hashtag Takeover. *Feminist Media Studies, 14*(6), 1105–1107. https://doi.org/10.1080/14680777.2014.975450.

Jaffe, S. (2018). The Collective Power of #MeToo. *Dissent, 65*(2), 80–87. https://doi.org/10.1353/dss.2018.0031.

Jane, E. A. (2014). 'Back to the Kitchen, Cunt': Speaking the Unspeakable About Online Misogyny. *Continuum, 28*(4), 558–570. https://doi.org/10.1080/10304312.2014.924479.

Jarrett, K. (2017). *Feminism, Labour and Digital Media: The Digital Housewife*. London: Routledge.

Jons, R. R. (2013). Is the "4th Wave" Digital? *Bluestockings Magazine*, 5 December 2019. Retrieved February 10, from http://bluestockingsmag.com/2013/08/19/is-the-4thwave-of-feminism-digital/.

Keller, J., Mendes, K., & Ringrose, J. (2018). Speaking 'Unspeakable Things': Documenting Digital Feminist Responses to Rape Culture. *Journal of Gender Studies, 27*(1), 22–36. https://doi.org/10.1080/09589236.2016.1211511.

Kim, Y. M. (2012). The Shifting Sands of Citizenship: Toward a Model of the Citizenry in Life Politics. *The Annals of the American Academy of Political and Social Science, 644*, 147–158.

López, E. (2018). Femicide in Ciudad Juárez Is Enabled by the Regulation of Gender, Justice, and Production in Mexico. *LSE Latin America and Caribbean Blog*. Retrieved May 10, 2019, from http://eprints.lse.ac.uk/88766/1/latamcaribbean-2018-02-15-femicide-in-ciudad-juarez-is-enabled-by-the.pdf.

Lopez, Y. Z. 2018. ¡Ni Una Mas!: Sexual Assault. It's an Issue Everywhere. But Let's Focus on Mexico. *The Medium*. Retrieved April 8, 2019, from https://medium.com/gendered-violence/ni-una-mas-sexual-assault-its-an-issue-everywhere-but-let-s-focus-on-mexico-7be0183b0ba3.

Marcelle, G. M. (2006). A Feminist Agenda for Reducing the Gender Digital Divide. In E. M. Trauth (Ed.), *Encyclopedia of Gender and Information Technology* (pp. 329–334). Hershey, PA: Idea Group Reference.

McLean, J., & Maalsen, S. (2013). A Geography of Revitalised Feminism. *Geographical Research, 51*(3), 243–256. https://doi.org/10.1111/1745-5871.12023.

Mdembe, A. (2003). Necropolitics (L. Meintjes, Trans.). *Public Culture, 15*(1), 11–40.

Mendes, K. (2015). *SlutWalk*. Basingstoke: Palgrave Macmillan.

Mendes, K., Ringrose, J., & Keller, J. (2018). #MeToo and the Promise and Pitfalls of Challenging Rape Culture Through Digital Feminist Activism. *European Journal of Women's Studies, 25*(2), 236–246.

Milligan, L. (2018). I am That Girl. *4 Corners*. Retrieved July 20, 2018, from http://www.abc.net.au/news/2018-05-07/kings-cross-rape-case-that-put-consent-on-trial/9695858.

Mulligan, C., & Carter, L. (2018). NSW Attorney-General Calls for Review of Sexual Consent Laws Following Four Corners Program. *ABC News*. Retrieved March 4, 2019, from http://www.abc.net.au/news/2018-05-08/nsw-attorney-general-calls-for-review-of-sexual-consent-laws/9734988.

Munro, E. (2013). Feminism: A Fourth Wave? *Political Insight, 4*(2), 22–25.

Musgrave, H. (2016). *#NiUnaMenos: Femicide in Argentina* (Law School International Immersion Program Papers, No. 15). Retrieved June 10, 2019, from https://chicagounbound.uchicago.edu/cgi/viewcontent.cgi?article=1038&context=international_immersion_program_papers.

Nakamura, L. (2015). The Unwanted Labour of Social Media: Women of Colour Call Out Culture as Venture Community Management. *New Formations, 86*, 106–112.

Ngai, S. (2005). Ugly Feelings (Vol. 6). Cambridge, MA: Harvard University Press.

Pitman, E. (2018). Misogyny Is a Human Pyramid. *Meanjin (online)*. Retrieved online August 9, 2018, from https://meanjin.com.au/blog/misogyny-is-a-human-pyramid/.

Qiqing, L. (2018). Former Doctoral Student Accuses Professor of Sexual Misconduct. *Sixth Tone*. http://www.sixthtone.com/news/1001491/Former%20Doctoral%20Student%20Accuses%20Professor%20of%20Sexual%20Misconduct.

Regulska, J. (2018). The #MeToo Movement as a Global Learning Moment. *International Higher Education, 94*, 5–6.

Rodino-Colocino, M. (2018). Me too, #MeToo: Countering Cruelty with Empathy. *Communication and Critical/Cultural Studies, 15*(1), 96–100.

Salami, M. (2013, September 26). African Women Are Blazing a Feminist Trail—Why Aren't We Listening to Them? *The Guardian*. Retrieved August 1, 2018, from https://www.theguardian.com/commentisfree/2013/sep/23/african-women-rwanda-feminism.

Salami, M. (2014, May 5). The Coming of (Digital) Age: How African Feminists Are Using the Internet to Change Women's Lives. *GenderIT.org*. https://www.genderit.org/articles/coming-digital-age-how-african-feminists-are-using-internet-change-womens-lives.

Sanchez, L., & Thomson, E. (1997). Becoming Mothers and Fathers: Parenthood, Gender, and the Division of Labor. *Gender & Society, 11*(6), 747–772.

Sills, S., Pickens, C., Beach, K., Jones, L., Calder-Dawe, O., Benton-Greig, P., et al. (2016). Rape Culture and Social Media: Young Critics and a Feminist Counterpublic. *Feminist Media Studies, 16*(6), 935–951. https://doi.org/10.1080/14680777.2015.1137962.

Stewart, K. (2007). *Ordinary Affects*. Durham: Duke University Press.

Subramaniam, S. (2014). *From the Streets to the Web: Feminist Activism on Social Media*. MPhil thesis, Tata Institute of Social Sciences, Mumbai. http://cscs.res.in/dataarchive/textfiles/from-the-streets-to-the-web-feminist-activism-on-social-media-sujatha-subramanian-tata-institute-of-social-sciences.

Sullivan, C. T. (2018). Indigenous Australian Women's Colonial Sexual Intimacies: Positioning Indigenous Women's Agency. *Culture, Health & Sexuality, 20*(4), 397–410. https://doi.org/10.1080/13691058.2017.1349930.

Tagny, G. (2016). Feminism Online in West and Central Africa: Identities and Digital Colonisation. Retrieved August 1, 2018, from https://www.genderit.org/articles/feminism-online-west-and-central-africa-identities-and-digital-colonisation.

Van Dijk, J. A. G. M. (2006). Digital Divide Research, Achievements and Shortcomings. *Poetics, 34*(4), 221–235.

Williams, J. E. (2015). Rape Culture. In G. Ritzer (Ed.), *The Blackwell Encyclopedia of Sociology*. Retrieved May 6, 2019, from https://doi.org/10.1002/9781405165518.wbeosr019.pub2.

Wright, M. W. (2011). Necropolitics, Narcopolitics, and Femicide: Gendered Violence on the Mexico-US Border. *Signs: Journal of Women in Culture and Society, 36*(3), 707–731.

Zeng, M. J. (2018). From #MeToo to #RiceBunny: How Social Media Users Are Campaigning in China. *The Conversation*. Retrieved July 20, 2018, from https://theconversation.com/from-metoo-to-ricebunny-how-social-media-users-are-campaigning-in-china-90860.

# 10

# Australian Feminist Digital Activism

Digital feminisms are strong and diverse. Within Australia, digital feminist practices have continued to emerge to address multiple forms of injustices. This chapter presents aspects of Australian digital feminism that aim to reduce sexism and misogyny in digital and non-digital spaces. Australian digital feminist spaces encompass a range of initiatives, including activist groups in social media, chatbots to support women trying to leave abusive relationships, and podcasts (for example Firth et al 2018). First, I introduce some of the major issues in sexism and misogyny that are the focus of Australian digital feminism and then give some insights into how creative use of digital tools are supporting these interventions.

The more-than-real can generate productive and destructive interventions, depending on the context and intentions of those engaging with digital technologies, and the affordances of digital spaces. In these ways, the more-than-real now provides more and different opportunities for sexist and misogynistic acts, including image-based abuse (frequently still referred to as revenge porn) where people can circulate intimate materials post-relationship as a form of revenge (Farries and Sturm 2018). These are serious and worrying issues that are being addressed by new laws and regulations on digital spaces to prosecute


J. McLean, *Changing Digital Geographies*,
https://doi.org/10.1007/978-3-030-28307-0_10

perpetrators in some jurisdictions, including in Australia and parts of the United States (Franks 2015). Legislatures in Malawi and Uganda have instituted anti-obscenity laws and anti-pornography laws but nothing specifically against image-based abuse (Chisala-Tempelhoff and Kirya 2016). At the same time, digital spaces are allowing people, institutions and communities to challenge these possibilities of abuse and violence and counter sexism and misogyny, among other damaging acts.

Refiguring the more-than-real through constructive interventions, feminist digital activism provides an opening for glimpsing change in social processes. Farries and Sturm (2018) define digital action that seeks remedies for injustices as 'cyberspatial justice' and suggest that this is limited, and needs supplementation through other measures such as copyright law, especially in the case of image-based abuse. They quote Franks (2011, 255) who argued that cyberspace is a number of places that serve to reinforce and effectively produce 'women's unwilling embodiment'. While this is true, it is also a space where women, and others, are *willingly* embodied and demonstrate agency through different avenues. Yes, structural disadvantages including sexism are reproduced online but they are also challenged. This is one of the key facets of the more-than-real: polarised possibilities of self-expression and assertion of identity sit alongside exploitation and domination of different forms.

## Multiple Feminisms, Multiple Forms

Just as there is no singular feminism, there is no singular Australian digital feminism. From spontaneous hashtag-driven social media campaigns, to organised non-governmental organisations, to startups providing tools for women to leave abusive relationships, to government-funded safety-focused organisations, Australian digital feminism is diverse and growing: Table 10.1 shows just a small sample of this diversity. The strategies that these interventions draw on are similarly diverse, including informal activist tactics, law reform and education and awareness, mobilising different approaches as appropriate to combat gender-based violence, everyday sexism, sexual harassment and image-based abuse.

**Table 10.1** A sample of Australian digital feminisms

| Institution | Date established | Funding | Remit | Areas of current campaign focus | Scale of campaigns |
|---|---|---|---|---|---|
| Destroy the Joint | 2012 | None | Highlight instances of sexism and misogyny | Gender-based violence, everyday sexism, political representation of women | Multiple: individual, communities, state-based issues (e.g. abortion laws), national, international (e.g. Women's March Washington) |
| Hello Cass | 2018 | Not-for-profit | Provide a medium for people experiencing abuse to get help | Gender-based violence | Individual and communities |
| eSafety Women | 2016 | Australian government | Help women experiencing family and/or domestic violence | Gender-based violence, image-based abuse, trolling | Multiple: individual, communities, state-based issues (e.g. education campaigns), national (e.g. law reform), international |
| eSafety Commission's WITS | 2018 | Australian government | Women Influencing Tech Spaces - increasing profile of women in digital spaces | #womenwithWITS | Multiple: individual, communities, social media sites |

Some efforts in Australian digital feminism are ephemeral while others are longer-lasting, usually linked to the motivations, capacities and interests of the collective that supports the emergence of specific efforts. There is a form of commodification occurring within digital spaces of feminist politics (Kenny 2017) as feminist philosophies and practices are co-opted for mainstream media sites like Mamamia (mamamia.com.au) and DailyLife (dailylife.com.au). Kenny (2017) argues that the original feminist slogan 'the personal is the political' should be reworked to 'the personal is the neoliberal'. However, while these processes are at play, there are other efforts that are not based in profit-generation that are forwarding feminist goals. Further, it is worthwhile considering the way digital networks are facilitating the telling of different stories in digital spaces outside of profit-generating objectives.

For example, Destroy the Joint (DTJ) is an Australian digital feminist group that arose from a hashtag moment in 2012 as a critical response to a conservative radio host, Alan Jones, who had declared that women are problematic in 'the joint': 'Women are destroying the joint – Cynthia Nixon in Melbourne, Clover Moore here. Honestly' (Farr 2012, paragraph 5). DTJ now works as an online social movement that has grown from a hashtag and diffuse conversation to become a targeted, curated space. Initially, the DTJ hashtag was a digital space or meeting point for people reflecting on the absurdity of Jones' claims that women in political life were destructive forces because of their gender, and grew to encompass critiques of gender inequality and lampooning sexist and misogynistic acts (McLean and Maalsen 2013; Gleeson 2016). People started to tweet the diverse ways in which women are apparently destroying the joint with the hashtags #jointdestroyer and #destroythejoint. Soon after, Jenna Price, a journalist and media academic helped establish DTJ as a social media movement on Facebook and Twitter (McLean and Maalsen 2013). As of 2018, DTJ has over 94,000 Facebook followers and 22,000 Twitter followers. A team of anonymous volunteers moderate DTJ, who collectively share a passion to stop sexism and misogyny (Gleeson 2016). The emergence of DTJ was made possible partly by the political climate in Australia at the time, and a wider discontent with gender inequality. As DTJ emerged, Australia's first female Prime Minister, Julia Gillard, was experiencing frequent sexism, from the general public, political commentators and politicians alike (Sawer 2013).

Somewhat differently to Fotopoulou's (2016) analysis of women's organisations that have existed prior to digital networks and then use these spaces to expand and/or complement their activism, DTJ arose as a result of the affordances of the digital. So rather than adapting to the digital, DTJ and other institutions that have arisen from the digital have slightly different challenges, including being taken seriously in academic and activist spaces and sustaining their work in corporate structures.

Campaigns by Australian digital feminisms are directed at ending gender and family-based violence, sexual harassment and everyday sexism. The strong focus on gender- and family-based violence in Australian digital feminism includes a number of strategies, ranging from recording and documenting the lives of those lost to such violence to campaigning for law reform and raising awareness of the issue. One such campaign is 'Counting Dead Women' by DTJ, a public record of instances of femicide since 2014 in Australia. The data is presented simply and clearly on social media sites run by DTJ (see Fig. 10.1).

In partnership with DTJ moderators, Cullen et al. (2018) have analysed the circumstances of women who were killed from January to December 2014 to identify risk factors, establish whether any patterns exist between these incidents, and what could be done to prevent femicide. The sheer number of women killed during the first two years of the count is staggering:

> Between January 2014 and December 2016, CDWA counted 234 femicides; approximating to one woman being violently murdered every 4.7 days in Australia across that time period. (Cullen et al. 2018, paragraph 4)

The circumstances of these murders were analysed by Cullen et al. (2018), drawing on publically available data on the victim, perpetrator and incident data, and using the CDW data as a baseline starting point. The research found that women are most at risk of fatal violence in their own home, committed by someone known to them, who is likely to be male, and this perpetrator usually had a history of violence. Further

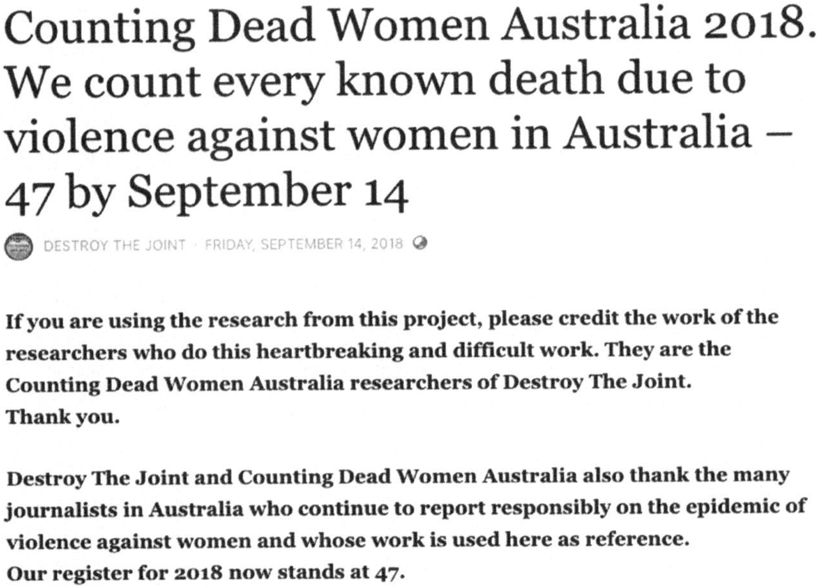

Counting Dead Women Australia 2018. We count every known death due to violence against women in Australia – 47 by September 14

DESTROY THE JOINT · FRIDAY, SEPTEMBER 14, 2018

**If you are using the research from this project, please credit the work of the researchers who do this heartbreaking and difficult work. They are the Counting Dead Women Australia researchers of Destroy The Joint.**
**Thank you.**

**Destroy The Joint and Counting Dead Women Australia also thank the many journalists in Australia who continue to report responsibly on the epidemic of violence against women and whose work is used here as reference.**
**Our register for 2018 now stands at 47.**

**Fig. 10.1** Screenshot of Counting Dead Women campaign page on Facebook

research is needed to establish the reasons for histories of intimate and family-based violence to escalate to murder (Cullen et al. 2018).

The digital record of femicides in Australia continues to grow each year and the activism that maintains this account is impressive and depressing; impressive in that people are able to continue recording such harrowing stories, and depressing that it has to keep being done. The CDW record forms the headline banner of DTJ at the moment (see Fig. 10.2) and the images of bowed heads of grieving women accompany the shocking tally. Here, the way digital activism generates affect—a growing tally and an annual total of femicides—reflects the affordances of the more-than-real.

In this digital space, records of these femicides are given context, the cumulative personal and familial costs are writ large, and the gendered nature of the crimes brought into stark relief. The combination of data and emotion is powerful and is helping to get information about these intolerable crimes more widely circulated. It is difficult to establish

**Fig. 10.2** Facebook page for DTJ—screenshot of banner

direct outcomes from this sort of work but there are some indicators of change in governance on these issues. For instance, in New South Wales a ministerial role for the Prevention of Domestic Violence and Sexual Assault was created in 2015 (Olding 2015). At the same time, the number of women who are dying in domestic situations is not reducing and the heavy work of agitating for social, legal and cultural changes in this area persists. In an interview with one of the moderators and leaders of DTJ, it was clear that this sort of work has a significant impact on feminists:

> But activism is different because you are trying to save women's lives or trying to improve women's existence. Um, I am really – I get very very bound up in trying to do that, and so it's got big emotional costs. Some weeks we have three deaths. And those are the weeks where I'm thinking "fuck, I can't do this anymore." I just feel miserable. (Ana, moderator DTJ interview, June 2018)

Affect and emotion play key roles in DTJ's activism; exhaustion, sadness, hope in trying to improve situations, all combine to underline a passion for doing more digital feminism. This corroborates with Gleeson's (2016) research on the all-consuming, and contradictory, qualities of digital activism and Anna-Britt's (2015) research on feminists in Ecuador and Peru. Not only are the boundaries of work extending beyond the usual business hours, also the definition of what work is, and can be, has shifted. Emotional labour goes into the creation, maintenance and control of these activist efforts as digital spaces are 'open' all day and night, and so too is the possibility of gaining and keeping the attention and support of those seeking alternative futures.

The participants of digital feminist spaces tend to draw on emotional motivations and intellectual arguments when contributing to interventions or debates in the line of their activist work. The more-than-real in particular facilitates such entanglement as issue publics often rely on drawing out personal motivations to drive public actions. In research co-written with Soph Maalsen and Sarah Prebble, published in *Gender Place and Culture*, we analysed three months' worth of Facebook conversations on DTJ to ascertain what were the main activist interests of DTJ participants and what drove such engagement. That research described how intersectional this digital activist space is, and highlighted the main concerns of those actively engaging in dialogue on DTJ over this time period. An example of the intersectionality of feminists working within DTJ follows:

> maybe we should concentrate on our own country first… we cannot be Marching in Solidarity against trump.. while we still have courts that keep releasing violent DV offenders back on the street to kill… when we still have Universities promoting a rapist of an indigenous woman, to a position over indigenous people….. when we have family law courts releasing children into care of men who are known or suspected of abuse…. When still women are dying at 1 a week in australia (sic)….. I just feel the more we spread ourselves out the less effect we have here… (8 likes). (McLean et al. 2019, 11)

This quote from a DTJ participant is particularly indicative of the broad and globally oriented approach of many Australian digital feminists. In listing a number of concerns about sexism and misogyny in Australia when responding to an original post from DTJ about the Women's March in the United States of Australia, this participant shared a sense of frustration at the number of issues that need urgent attention and a sort of poetic expression with each line leading with 'when…' The breadth and seriousness of interventions that spaces like DTJ offer is clearly at times overwhelming, for participants and moderators alike.

Further, intersectional concerns are evident in this DTJ participant's perspective, something that is widespread throughout the feminist action found in that space.

## Grounding Digital Spaces

The assumption that digital activism is frequently meaningless, captured in commentary from Gladwell (2010) and McCafferty (2011), misses the point of the work that is emerging from these spaces, and usually fails to closely analyse what is occurring in these spaces, let alone gaining insights from talking with those individuals, communities and institutions that use digital spaces. For instance, McCafferty (2011, 18) states that 'In the end, activism has always been—and will always be—about people. Specifically, people who show up in person'. This embodied presence argument suggests that 'showing up in person' does not happen in digital spaces and reinforces claims that digital spaces are lesser, or not as substantive, as face-to-face contexts. Connected with this immaterial conceptualisation of digital spaces is a sense that digital activism is superficial or secondary to the 'real' work of protesting in traditional contexts, such as rallies or door-to-door petitioning.

In contrast to McCafferty's argument, research on social media use for activism to stop sexual violence in United States higher education contexts shows that activists do not tend to separate social media activism and activism (Linder et al. 2016). Rather, different modes of activism are closely interrelated. Using ethnographic research techniques, Linder et al. (2016) interviewed 23 activists who reported that social media was an important tool to engage people in activism—it helps raise awareness, transforms power relations and builds community. From research that I have undertaken for this book, it is clear that this resonates with Australian digital feminisms. Activists and activist groups are acutely aware of the strengths and limitations of digital action and have sophisticated strategies to work effectively in compromised spaces.

One of the moderators of DTJ reflected on how digital action provides a good platform for engaging people and building solidarity, but is also dull and challenging:

*Jess*: How sustainable do you think online action is?
*Ana*: Okay. So when people ask me that question I often take out the online. And recognize that activism is activism. I don't do that by myself, oh fuck, you need a kind of army really. And you know, recruiting people who want to do this, that's hard…whenever anyone very young has become a moderator, they don't last very long, because activism is not exciting.

Collapsing the binary between digital and non-digital, online and offline, real and not real is an important strategy for this feminist activist, following a long tradition of feminist critiques that aim to trouble the binary thinking that lessens the second pair of the duo (Gibson-Graham 2011; Rose 1993).

The work of reducing sexism and misogyny in digital spaces does not exclusively belong to not-for-profit activists in Australia. Startups and federal government institutions are also trying to promote gender equality in a number of ways. For example, Hello Cass is a startup initiative that is offering a safe space for women experiencing domestic violence to talk about their experiences and get support. Hello Cass is a chatbot that is intended to offer women the first point of contact to help them getting support. The Founder/Director of Hello Cass describes the digital tool as a bridge to getting further assistance if required:

> And then I … looked at some research that came out of the States around mental health chatbots and they did some work around anxiety and depression, and these virtual therapists, and the findings were that people were actually more likely to be really honest and give more detail when they're talking to a simulated therapist than they were when they were talking to an actual human therapist. And it's because of shame and stigma and also, when we talk about things, when we disclose things, we take on the other person's reaction, as well, so we kind of want to protect them a little bit from our own pain. So the idea with Cass was if we made it really easy to maybe make that first disclosure and give a little bit of encouragement and just a little bit of information, then that might be enough to empower the user to kind of then… They can feel confident that they've got the right phone number to start a conversation with a person. (Founder/Director Hello Cass September interview, October 2018)

The Founder/Director of Hello Cass highlights the emotional barriers to full disclosure of dangerous situations and how they have created a medium that allows those experiencing domestic violence to seek support in a safe way. The hybrid entity of the chatbot—one that hovers in the more-than-real—is a bridge between suffering alone and getting appropriate help. Shame and stigma are not such important issues when talking to something that is more-than-real, they are emotions that can be put aside, and/or carried with less concern about judgement, in that exchange. The Founder/Director of Hello Cass went on to say that once someone who is experiencing abuse from a partner has talked about it in at least one instance, then they can talk about it again with someone who can help them, and start the process of becoming safe.

In Australia, there are government-sponsored initiatives that are aimed to increase digital safety that could be read as feminist. According to the eSafety Commissioner, Julie Inman Grant, the incidents that precipitated the formation of the eSafety Office and the establishment of the Enhancing Online Safety Act of 2015 came about in the two years before that legislation with a number of serious events. In 2014, public awareness of digital violence was raised with 'the tragic death of Charlotte Dawson…she was trolled and had a nervous breakdown and was trolled again and ended up committing suicide' (eSafety Commissioner interview, September 2018).

In 2013 another victim of cyberbullying, Chloe Fergusson, took her life at the age of 15. Chloe had experienced years of bullying in her Tasmanian school and suicided a few days after being physically attacked, having this abuse filmed, and was facing threats of this footage being circulated on social media. A petition led by family, friends and supporters called 'Chloe's Voice' gathered 49,000 signatures and was tabled in Federal Parliament in September 2014 by Senator Eric Abetz (Youth Network of Tasmania 2014). The coalition government then committed to creating a legislative framework to establish the eSafety Commissioner role and office and actively intervene in these sorts of abuse. The public attention on these worrying incidents translated into petitions, government support and funds to establish a new institution, initially with the brief to protect children, which was expanded, less than a year later, to all Australians. Australia is the first country in the world to have a designated eSafety Commissioner role and other nations are approaching the institution to find out more about their work, and to determine how they could set up similar institutions (eSafety Commissioner, pers comm).

We know that digital spaces are enablers of abuse as well as facilitators of help and support. The eSafety Women campaign instigated by the eSafety Commission draws attention to the broad gamut of ways in which women are subject to technology-enabled abuse and aims to provide practical measures to stop this happening. According to the eSafety Commissioner:

> eSafety women is a program that's directed at frontline workers working with women who've experienced family or domestic violence. We know that in 98% of family and domestic violence cases, particularly when a woman and her children are trying to escape that situation, technology facilitated that abuse. It's an extension of that abuse. We also know particularly when a woman's escaping that kind of situation, her phone or technology becomes a lifeline so that she can be in touch with social services, police if she needs to be, but also close friends and family members.
>
> But what social workers and the women themselves were finding is that it was opening them up. It was making them vulnerable to the stalking surveillance. Again, I guess furthering that abuse. This program, we've [the eSafety Office] got extensive resources on our website about checklist on how to make sure that if you've moved interstate and you're going to change a bank account that, your bank account records aren't going to your former partner. Ensuring that the phone or the iPad that the ex-partner gives to your kids doesn't have spyware on it.

The double-edged sword of digital technologies is important with respect to the eSafety Women campaign. Giving clear, practical information on how to protect women and children in times of high conflict and post-separation is instrumental to ensuring that the damage inflicted by digital tools is countered or ameliorated by practical interventions that protect people, and enable safety. It is much more than raising awareness that sits at the heart of the eSafety Women campaign: measures to empower women and to counter abusive behaviour are outlined in straightforward terms.

Another campaign that the eSafety Commission is running aims to proactively transform social media spaces. It is known that gendered digital abuse is widespread (Jane 2016) and that there is a pressing need to address such behaviour. The eSafety Commission provides a sort of flow-chart (https://www.esafety.gov.au/women-influencing-tech-spaces/taking-action#respond) to guide people through the best way to respond to cyber abuse (Fig. 10.3).

The eSafety Commission's clear framework for how to respond to cyber abuse contributes to the demystifying of these sorts of digital behaviours and builds a language around such acts. Further, it outlines what support the Commission and other institutions offer to stop such abuse. Rather than suffering alone or pursuing what Jane (2016) calls feminist digilantism, this is a government measure that can help produce better digital spaces. The eSafety Commission's work includes tips on how to insulate people from cyber abuse under the following invitation: 'Let's build your psychological armour'. The collective is emphasised again here, rather than the individual who may be exposed to harm, and responsibility is distributed rather than carried alone. I read this as an example of what Jane (2016, 284) argues is much needed in seeking safer digital spaces. Jane (2016) states that feminist digilantism

**Fig. 10.3** eSafety Commission publication 'Skills and strategies for coping with cyber abuse'

is limiting and that 'a hybrid of feminist activist efforts – including a recalibrated approach to collectivism – is required to achieve the legislative and corporate reforms necessary to address the significant social problem of gendered hate on the Internet'. We could read the eSafety Commission work as a step towards the necessary reforms that Jane suggests; the eSafety Commission is a product of Federal Government legislation and is an institution set up to specifically address serious cyber abuse.

## Creative Uses of Digital Spaces: Emotion and Affect in Feminist More-Than-Real Spaces

The personal action frames that are mobilised in Australian digital feminisms are shaped by diversity and intersectionality. Similarly, Bennett and Segerberg (2012) describe how occupy's slogan 'we are the 99 per cent', that arose from the United States origins of the movement, was inclusive of different personal reasons to contest structures that need changing. The openness of the political goals of *occupy* led to its global networking. In an analogous way, DTJ's slogan as of 2018—'demand a better world for women and girls'—is a porous claim that is adaptable to particular contexts. The work of the eSafety Commission is also broadly defined: 'The Office is committed to helping all Australians have safe, positive experiences online' (eSafety Commission 2018). Each of these claims invoke emotive frames by using terms such as 'demand' and 'positive', while empowerment may accompany realisations of being a part of the 99%. Activist efforts in these contexts combine rational and affectual elements, and connective and collective action. Bennett and Segerberg (2012, 745) ascribe a 'boundary zone' where connective action gives way to collective action in their analysis of the logic of connective action exegesis but later describe how these modes of action are overlapping and entangled. I think that this is an important point for understanding how the more-than-real works as the separation between individual and collective goals is very rarely clear in the issue publics that can emerge in social media. Indeed, affectual elements are usually central to making sure digitally networked action *works* (Gerbaudo 2012).

Government-funded initiatives, such as the eSafety Commission, are aware of the power of social media to 'give voice'. From the eSafety Commissioner:

> I do think social media has proven to be a very powerful tool for sparking and sustaining online movements. Really the whole idea of Twitter is speaking truth to power and I think the Arab Spring and #Illridewithyou and #MeToo, and all of these movements would not have achieved, I guess the same kind of vitality if we were only relying on the news outlets almost as a filter. Because when you think about it, they are much more subjective vis-a-vis say social media where everyday people can extensively have an equal share of voice at least on those respective platforms.

Vitality stems from the personal connections and narratives that enable these digital moments. An important qualification is required here as 'speaking truth to power' is possible in social media but it often does not translate to being listened to, due to inequitable structures that are reproduced in the digital, and significant issues with attention fatigue (Zuckerman 2013). Digital spaces can enable you to speak, yet they do not guarantee that you will be heard or attended to. Further, there is sometimes a clustering effect in digital spaces that can lead to networks being quite concentrated.

There is nothing inherently wrong with this concentration, but it is important to point out the limitations involved in some attempts at transforming the digital. The creative use of digital tools to support feminist interventions is sometimes overlooked when analysing how social and cultural change is facilitated. The digital feminisms that are the focus of this chapter all draw on digital technologies in different ways and share similar goals. As introduced earlier in this chapter, the name 'Destroy the Joint' came from a reclamation of a term that a renowned sexist commentator applied to women in politics. Twitter users inverted the insult and a movement grew from a moment of parody as people shared the thousands of ways that women are constantly building rather than destroying any vaguely defined joint. Play was central, then, to a serious endeavour to overturn sexism and misogyny. The personal politics that drive DTJ account for some of its success, as does its unified but not uniform structure (Mclean et al. 2016). Paradoxes abound in these digital spaces.

Play can produce alternative worlds and its potential to transform lies in the ability for participants to identify what is playful and what is not. Social and cultural conditioning leads to this capacity and Johnson's (2017) research on parody in Twitter illustrates how play allows users to work within and around platforms, using their own agency. Drawing on Bakhtin's analysis of play, Johnson states that 'play thus allows us to move between different possible worlds and lives' (Bakhtin 1968 [1965]; Johnson 2017, 26).

Play can invoke a form of prefigurative politics with a mirroring of means and ends, as the change people want to see in the world is created through the acts to achieve that change. Facilitating platforms that have democratic aspects to them is a form of prefigurative politics in the sense that it is 'a political action, practice, movement, moment or development in which certain political ideals are experimentally actualised in the "here and now", rather than hoped to be realised in a distant future' (van de Sande 2013, 230). Australian digital feminist movements and spaces critique patriarchal power in remaking their own safe spaces. The process, then, is as important as the outcome achieved. Here, the performative elements of play within online activism come to light for the subversive aspects of disruptive play are aligned with the goals of equality that underpin feminist and diversity activism.

Corporate spaces such as Facebook and Twitter are remade in decidedly not-for-profit initiatives and we could read DTJ as an example of a hybrid research collective (Gibson-Graham 2011), as assemblages of human and non-human elements that come together in an attempt to make different worlds, by themselves *working as* different worlds. Further, all three cases so far mentioned in this chapter—Hello Cass, DTJ and the eSafety Commission—operate without a profit imperative and are an example of the diverse economies (Gibson-Graham 2008) relying on volunteer contributions, government funding or the support of a parent corporation to underpin their existence. They all exploit pre-existing platforms for feminist ends. In this way, there is a creativity that conjoins these efforts of deliberate interventions that remakes these spaces that are also deeply problematic. For example, Bivens (2017) analyses Facebook over a ten year period and concludes that Facebook's software and its users are structured by a gender binary, despite superficial efforts to afford gender diversity. The structure of software that makes Facebook includes deeply structured gender binaries but Bivens (2017) also points out that gender hacking occurs as users push back against these structures.

Taking seriously this diversity and the contradictory possibilities of the digital means that I do not seek to reduce the complexities that are inherent to relatively open social media space (Segerberg and Bennett 2011) but that I seek to better understand the practices, technologies and subjectivities that co-produce digital feminisms. The podcast 'After #MeToo: Stories of social change' (Firth et al 2018), a collaboration by 2SER, a community radio station, and the Centre for Social Justice and Inclusion at the University of Technology Sydney, is another example of digital feminism in the Australian context. It is hosted by Verity Firth, head of that Centre, and produced by radio journalists Ollie Henderson, Miles Herbert and Ninah Kopel. The podcast has covered stories of immigrant women navigating hostile court systems as survivors of domestic violence and women's safety in public places. The use of a story-based structure for the podcast allowed for multiple ways of knowing to exist alongside each other. As Griffiths (2007) states, 'stories change the way people act, the way they use available knowledge. The stories we live by determine the future. So, in harnessing the power of narrative, in listening to, rediscovering and generating true stories, we change the world'. So by sharing these stories in an accessible podcast format, social transformations that are feminist informed may be further encouraged.

The roles that emotion and affect play in digital activist endeavours are intangible but important. For example, Kwan (2007) asks geographers to consider feminist interventions in cultural spaces. Her analysis of GIS-based art projects, one of which was her own creation, demonstrates how technologies can rework emotional geographies. She concluded that if we want 'to effect broader social change… it is important to scale our care or concern from the personal/local level up to larger contexts' (Kwan 2007, 29). The rise of feminist digital movements such as DTJ is one such instance of a larger context where multiple scales of feminist action are pursued. What emerges from an analysis of the different sorts of feminism emerging from digital spaces is the liminal and paradoxical dimensions of human–technology relations that underpin change (Kwan 2007). The work of the eSafety Commission has the potential to reach many as well, given its government funding and standing.

## Sustaining Australian Digital Feminisms

Feminists have long challenged binary thinking and brought together, rather than separated, ways of knowing. Following Bennett's (2010) vibrant matter, this analysis of Australian digital feminist actions seeks to also highlight what is often left in the shadows in anthropocentric readings of technologies to consider the material agency of non-human or 'not-quite-human things' (Bennett 2010, 10). When interviewing people involved in making and wrangling these digital spaces, questions about the sustainability of these actions stood as proxy to considering the agency of these engagements. I asked research participants whether their digital action was sustainable, economically and environmentally, and complex responses were offered. Feminist activists are conscious of the work of more-than-humans in connection to the goals that they seek to achieve. Clearly, as Duffy et al. (2018, 2) find in research on affect and sustainability leaders, 'They (the practices and performances that sustainability leaders adopt) involve the enrolment of non-human forces alongside sets of ideas that structure what people do'.

Activism in digital spaces is feasible partly due to the convenience of the technologies. When asked about the sustainability of digital activism, a DTJ moderator said that:

> 'Activism is not exciting, except on the odd occasion when you have a victory. And that, I mean how many times a year does that happen? Once or twice? Not very often. So is activism sustainable? Yes, because we recognize that there are, um, moves in equality and moves in justice. Um, how it's sustained is depending on the weather, the people, the action. I- I'm lucky to be in a big group, 'cause if I was still doing this with an isolated person with a photocopier that would be freaking miserable.' (Ana, DTJ moderator interviewed, June 2018)

The contingency of persistent activism is due to random variables—the weather (humorously), people, 'the action'. Also, the size of the group sharing the busy and relentless work of moderation is crucial. Last, the vibrant matter—digital spaces—that enables the connections between moderators to produce an effective activist space plays a significant role. This digital feminist draws out the contrast between traditional activism (with its miserable photocopier) and implies the immediacy and deep connectivity facilitated by the digital. The agency of this non-human tool can be seen here.

The language used to depict digital spaces and issues within them draws on environmental metaphors. The eSafety Commissioner discussed sustainability of digital spaces in terms of environmental contamination and responsibility.

> I think we're starting to talk about platforms in terms of the level of toxicity, a pollution. How much meaningful dialogue is there versus all the abuse or the negativity or even if it's the ads or the spam. I think we should think of online spaces the way we do environmental spaces. The platforms themselves have a responsibility for keeping those cleaner. One of the things that the initiatives that I've started that I feel that's starting to get some legs, both here and globally, is an effort to encourage safety by design. (eSafety Commissioner, September 2018)

Conceptualising platforms as having contamination that needs remediation is interesting as it reframes these digital spaces as having a physicality that needs refiguring. The eSafety Commissioner goes on to discuss how safety by design is something that should be introduced across the board, from small digital initiatives to the dominant platforms such as Amazon and Google.

In a quite different way, Hello Cass Founder/Director recognised the vibrancy of digital entities and the requirement to manage these. She described the startup in the following way:

> 'So it's expensive because it's like it needs constant cultivation. It's a living kind of thing. So you need to be constantly working on it. And that's where the cost is. The challenge is, I think, with community-based things

> and not-for-profits is that they can't really afford, at the moment, to have full-time technical practitioners working'.

The agency of non-human, or digital, things comes to light in this quote as the Hello Cass Founder/Director has worked in technological companies for over a decade and knows the depths and complexities of digital geographies. They know that the maintenance of such digital things involves constant tending too—there is no 'set and forget' if you want a digital thing to keep working.

Referring to the Human Rights and Technology Conference that was held in July 2018 in Sydney Australia, the Hello Cass Founder/Director discussed concerning gaps in dialogues about making better digital things:

> This is something that I worry about… And I've worked for big tech companies that have huge emissions output. Emissions that they produce, emissions that are next to their name just with their data centres and things like that. So the fact that we don't have any conversation about green tech or greening tech or even the fact that to use the internet, there's a cost – it's a real worry.

Framing environmental dilemmas as social and cultural dilemmas emerges here, as does the weight of those. Data centres are framed by Fish and Garrett (2019) as 'byte bunkers' that afford desire for endless consumption of digital worlds, concrete structures that provide the base and housing of more-than-real lives. A different way of knowing digital spaces must play a role in belonging in this new mode of humanity, extending Gibson-Graham's (2011) challenge to rethink connection. Enacting new modes of feminist living can incorporate digital sustainability where the environmental costs and benefits of digital things are fully incorporated in how we figure our daily lives.

Anthropocene feminisms have offered readings of the always-entangled human and more-than-human worlds in light of the discussions surrounding the 'new' geological epoch. Colebrook (2018, 18) argues that 'the Anthropocene is the counter-Anthropocene'. The recognition that centuries of environmental, human and political exploitation has produced geologically significant change is also the inflection point at which critiques of those actions appears. Humorously and tragically, Colebrook (2018, 18) exclaims:

> We look at the geological scar and remark to ourselves, as though something has changed: now, finally, the earth is telling us that we have impact. "My goodness, who would have thought that centuries of slavery, violence, kleptocracy, plundering and then liberation of some humans at the expense of others—who would have thought this was a destructive indictment of 'the human'? Who could possibly have imagined that our species was destructive of its milieu without the definitive evidence of the geological record?"

The patterns of violence and theft that have produced this geological moment are aligned with the same relations that oppress those of difference, and among these differences are gendered, sexuality, racial and class-based identities. The fact that a geologic marker is required, a geological scar, to verify this relationship gives Colebrook (2018) valid grounds for a moment of mockery. The exceptional thinking that underpins the shock and surprise at the Anthropocene is part of the problem. Colebrook (2018) suggests that the Anthropocene moment perpetuates exceptionalism as it provides evidence of a continuation of 'man' thinking that he can be distinct from nature, and above making a difference to that nature. The digital feminists that have shared their insights here show how connections between different forms of exploitation are already present in the work that they do, at least in some forms.

## Conclusion

Space is the realm of possibility and heterogeneity, as Massey (2005) argued in her spatial manifesto, and digital spaces are exemplars of this multiplicity. The empirical data that informs this chapter covers a range of digital spaces, from social media issue publics to podcasts to chatbots, and diverse institutional settings, including government, not-for-profits, startups and community broadcasters. There is diversity between different sorts of digital feminisms and diversity within particular digital communities. While media scholars have argued that feminist digilantism (Jane 2016) is limited and limiting, I have shown that there is already a place for this form of advocacy and that there is value in such grassroots-driven feminism, along with other researchers such as Gleeson (2016) and Cullen et al. (2018).

Recognition of the capacity for digital spaces to deliver misogyny and sexism as well as the affordances of the same to challenge, subvert and overturn such behaviours and attitudes is common to each of the cases discussed here. The more-than-real can have polar extremes in terms of the destructive and generative possibilities that it affords. Gender relations are not immune to these polarities but individuals and communities continue to pursue change from, and with, the digital in pragmatic ways. We could read some Australian digital feminists as techno-realists, rather than techno-dystopians or utopians, as those that I interviewed who are involved in making digital spaces safer are conscious that they are working with these possibilities. Part of this pragmatism includes recognition of the intersection of digital spaces with environmental realities, as demonstrated in conversations about sustainability of digital action. Metaphors of environmental pollution are used to interface with social and cultural safety while some digital feminists are also acutely aware of the carbon costs of digital action and how these are nearly universally externalised.

An attunement to emotion and affect continue to propel Australian digital feminism. The intermingling of logical arguments to stop sexism and misogyny with emotive appeals for action underlies some explicit feminist interventions in the digital. This is a seemingly never-ending task as digital technologies include affordances for relative safety and new opportunities for abuse or violence. Further, the environmental framings of digital geographies by those working on and with feminist digital spaces adds an inflection that integrates important qualities of the more-than-real. The work of making digital lives sustainable, in the many ways that this is defined, is not left to external others, or technologists and corporations who are enthusiastically remaking digital geographies. Rather, digital feminists are critiquing and engaging with digital spaces as parts of problematic modernist systems that are up for the reworking—for humans and nonhumans, and for diverse priorities. The surprising possibilities that emerge with more-than-real spaces are intriguing in the context of Australian digital feminisms. The way that chatbots, for example, may provide a novel bridge for those experiencing gendered violence to get out of fraught situations points to the generative possibilities of these technologies. Providing a way to get help that may directly overcome very human emotions of shame and stigma is a substantial strength of this digital technology.

## References

Anna-Britt, C. (2015). "I Am Not Just a Feminist Eight Hours a Day": Youth Gender Justice Activism in Ecuador and Peru. *Gender & Society, 29*(6), 888–913. https://doi.org/10.1177/0891243215599647.

Bakhtin, M. 1968[1965]. *Rabelais and His World* (Helene Iswolsky, Trans.). Cambridge, MA: MIT Press.

Bennett, J. (2010). *Vibrant Matter: A Political Ecology of Things*. Durham: Duke University Press.

Bennett, W. L., & Segerberg, A. (2012). The Logic of Connective Action: Digital Media and the Personalization of Contentious Politics. *Information, Communication & Society, 15*(5), 739–768.

Bivens, R. (2017). The Gender Binary Will Not Be Deprogrammed: Ten Years of Coding Gender on Facebook. *New Media & Society, 19*(6), 880–898.

Chisala-Tempelhoff, S., & Kirya, M. (2016). Gender, Law and Revenge Porn in Sub-Saharan Africa. *Palgrave Communications, 2*, article number 16069. Retrieved July 10, 2019, from https://www.nature.com/articles/palcomms201669.

Colebrook, C. (2018). We Have Always Been Post-Anthropocene: The Anthropocene Counterfactual. In R. Grusin (Ed.), *Anthropocene Feminism* (pp. 1–20). Minneapolis: University of Minnesota Press.

Cullen, P., Vaughan, G., Li, Z., Price, J., Yu, D., & Sullivan, E. (2018). Counting Dead Women in Australia: An In-Depth Case Review of Femicide. *Journal of Family Violence*. https://doi.org/10.1007/s10896-018-9963-6.

Duffy, M., Gallagher, M., & Waitt, G. (2018). Emotional and Affective Geographies of Sustainable Community Leadership: A Visceral Approach. *Geoforum*. https://doi.org/10.1016/j.geoforum.2018.09.005.

eSafety Commission. (2018). *About the Office*. Retrieved September 21, 2018. https://www.esafety.gov.au/about-the-office.

Farr, M. (2012, August 31). Alan Jones: Women Are Destroying the Joint. *The Herald Sun*, 2013. Retrieved March 15, 2018, from http://www.heraldsun.com.au/news/alan-jones-women-are-destroying-the-joint/news-story/402a32a90a61275ac789a6cbfb45c31d.

Farries, E., & Sturm, T. (2018). Feminist Legal Geographies of Intimate-Image Sexual Abuse: Using Copyright Logic to Combat the Unauthorized Distribution of Celebrity Intimate Images in Cyberspaces. *Environment and Planning A: Economy and Space*. https://doi.org/10.1177/0308518x18786964.

Firth, V., Ollie, H., Herbert, M., & Kopel, N. (2018). *After #MeToo: Stories of Social Change* [podcast]. Retrieved September 26, 2018, from https://player.whooshkaa.com/shows/after-metoo-stories-of-social-change.

Fish, A., & Garrett, B. (2019). Resurrection from Bunkers and Data Centers. *Culture Machine, 18*. http://culturemachine.net/wp-content/uploads/2019/04/FISH.pdf.

Fotopoulou, A. (2016). Digital and Networked by Default? Women's Organisations and the Social Imaginary of Networked Feminism. *New Media & Society, 18*(6), 989–1005.

Franks, M. A. (2011). Unwilling Avatars: Idealism and Discrimination in Cyberspace. *Columbia Journal of Gender & Law, 20*(2), 224–261.

Franks, M. A. (2015, August 17). *Drafting an Effective "Revenge Porn" Law: A Guide for Legislators*. Retrieved June 3, 2019, from http://dx.doi.org/10.2139/ssrn.2468823.

Gerbaudo, P. (2012). *Tweets and the Streets: Social Media and Contemporary Activism*. London: Pluto Press.

Gibson-Graham, J. K. (2008). Diverse Economies: Performative Practices for 'Other Worlds'. *Progress in Human Geography, 32*(5), 613–632.

Gibson-Graham, J. K. (2011). A Feminist Project of Belonging for the Anthropocene. *Gender, Place and Culture, 18*(1), 1–21. https://doi.org/10.1080/0966369X.2011.535295.

Gladwell, M. (2010). Small Change: The Revolution Will Not Be Tweeted. *The New Yorker*. Retrieved December 4, 2018, from https://www.newyorker.com/magazine/2010/10/04/small-change-malcolm-gladwell.

Gleeson, J. (2016). '(Not) Working 9–5': The Consequences of Contemporary Australian-Based Online Feminist Campaigns as Digital Labour. *Media International Australia, 161*(1), 77–85.

Griffiths, T. (2007). The Humanities and an Environmentally Sustainable Australia. *Australian Humanities Review, 43*. Retrieved February 4, 2019, from http://australianhumanitiesreview.org/2007/03/01/the-humanities-and-an-environmentally-sustainable-australia/.

Jane, E. (2016). Online Misogyny and Feminist Digilantism. *Continuum, 30*(3), 284–297. https://doi.org/10.1080/10304312.2016.1166560.

Johnson, A. (2017). *Twitter and the Body Parodic: Global Acts of Re-creation and Recreation*. Ph.D. thesis, MIT. Retrieved February 18, 2019, from https://dspace.mit.edu/handle/1721.1/113741.

Kenny, K. (2017, September 1). Popular Feminism in the Digital Age: How the Personal Has Been Made Commercial. *ABC News Opinion*. Retrieved September 14, 2018, from https://theconversation.com/friday-essay-the-personal-is-now-commercial-popular-feminism-online-79930.

Kwan, M. P. (2007). Affecting Geospatial Technologies: Toward a Feminist Politics of Emotion. *The Professional Geographer, 59*(1), 22–34.

Linder, C., Myers, J. S., Riggle, C., & Lacy, M. (2016). From Margins to Mainstream: Social Media as a Tool for Campus Sexual Violence Activism. *Journal of Diversity in Higher Education, 9*(3), 231–244.

Massey, D. (2005). *For Space*. London: Sage.

McCafferty, D. (2011). Activism vs. Slacktivism. *Communications of the ACM, 54,* 17–19. https://doi.org/10.1145/2043174.2043182.

McLean, J., & Maalsen, S. (2013). A Geography of Revitalised Feminism. *Geographical Research, 51*(3), 243–256. https://doi.org/10.1111/1745-5871.12023.

McLean, J., Maalsen, S., & Grech, A. (2016). Learning About Feminism in Digital Spaces: Online Methodologies and Participatory Mapping. *Australian Geographer, 47*(2), 157–177. https://doi.org/10.1080/00049182.2016.1138842.

McLean, J., Maalsen, S., & Prebble, S. (2019). A Feminist Perspective on Digital Geographies: Activism, Affect and Emotion, and Gendered Human-Technology Relations in Australia. *Gender, Place and Culture*. https://doi.org/10.1080/0966369X.2018.1555146.

Olding, R. (2015). New Anti-domestic Violence Minister Pru Goward Highlights Prevention. *Sydney Morning Herald*. Retrieved September 17, 2018, from https://www.smh.com.au/national/nsw/new-antidomestic-violence-minister-pru-goward-highlights-prevention-20150401-1mcvt3.html.

Rose, G. (1993). *Feminism & Geography: The Limits of Geographical Knowledge*. Minneapolis: University of Minnesota Press.

Sawer, M. (2013). Misogyny and Misrepresentation: Women in Australian Parliaments. *Political Science, 65*(1), 105–117. https://doi.org/10.1177/0032318713488316.

Segerberg, A., & Bennett, W. (2011). Social Media and the Organization of Collective Action: Using Twitter to Explore the Ecologies of Two Climate Change Protests. *The Communication Review, 14*(3), 197–215.https://doi.org/10.1080/10714421.2011.597250.

van de Sande, M. (2013). The Prefigurative Politics of Tahrir Square–An Alternative Perspective on the 2011 Revolutions. *Res Publica, 19*(3), 223–239.

Youth Network of Tasmania. (2014). Cyberbullying and Young Tasmanians. Retrieved August 1, 2018, from https://www.ynot.org.au/sites/all/files/ynot/docs/Cyberbullying%20and%20Tasmanian%20Young%20People_YNOT%20Report%20Dec%202014.pdf.

Zuckerman, E. (2013). *Rewire: Digital Cosmopolitans in the Age of Connection*. New York: W. W. Norton.

# 11

# 'It's Just Coding': Disability Activism In, and About, Digital Spaces

Digital technologies offer uneven affordances for different people, like other technologies. Particular devices and software are more accessible than others, while some technology companies are more responsive to calls for change than others. Digital activism by disability advocates has focused on reducing 'digital gaps' and calling corporations and governments to account for allowing and extending inaccessible and exclusionary tools. Such work is layered upon pre-existing campaigns for better support and access by disability advocates in diverse contexts, from the public to the private. Digital disability activists frequently use social media and formal governance processes to achieve social, technological and cultural changes, despite these spaces frequently making this work difficult.

Disability activists have forced manufacturers and designers to change digital technologies that are beneficial for all users, as universal design elements improve the functionality of a range of devices, tools and software. Ellis et al. (2015) discuss the constraints, issues and inaccessibility of digital technologies for many people with disabilities and identify the embedded qualities of digital spaces in everyday life. The entanglement of humans and digital technologies (O'Neal Irwin 2016)


J. McLean, *Changing Digital Geographies*,
https://doi.org/10.1007/978-3-030-28307-0_11

can cause friction, especially for people with disabilities if the digital is not designed with universal access in mind, and if neoliberal inclusionism dominates (Mitchell and Snyder 2015). So the practical usefulness of certain digital technologies for some groups and individuals in societies are simply not so, if they are not universally usable. Simultaneously, digital technologies can be inclusive if they provide new ways to navigate exclusionary spaces, such as removing the need to access difficult places.

This chapter draws on disability activists' perspectives to share narratives of how the more-than-real works from particular positionings, at times excluding people and communities, and at others facilitating calls for change. The contradictions of digital technologies are sharply delineated in this context as people with disabilities can be marginalised and/or gain greater independence from different digital interventions. 'Universal access' and 'universal design' are dominant discourses in disability activism and reformist approaches to digital technologies but also carry problematic conceptualisations. Within the critical disabilities studies literature, neoliberal inclusionism has been extensively critiqued, as it only allows certain kinds of people, and, in particular, certain kinds of bodies, to be included within social and economic structures, potentially at the expense of other marginalised peoples (Mitchell and Snyder 2015). The conditional acceptance of some people with disabilities through inclusion efforts is read as a biopolitical form of control, where people of difference are normalised for mainstream participation. Mitchell and Snyder (2015) argue that a crip/queer politics of difference is required to unsettle these processes of neoliberal inclusionism and allow for diversity to flourish. 'Crip' is used here in connection with crip theory, noting Kafer's (2013, 1) statement that claiming crip 'can be a way of acknowledging that we all have bodies and minds with shifting abilities, and wrestling with the political meanings and histories of such shifts'. I keep these critiques in mind in this chapter, especially when analysing how the structures of digital and non-digital spaces intersect with the politics and lived experiences of people with disabilities.

## Digital Activism by People with a Disability in Australia

Disability is approached in this chapter not as an individual, medical condition, but socially, culturally, politically and environmentally constructed, following the lead of legal theorists Steele and Thomas (2014). There are a number of disability activist groups in Australia which are working in and with digital spaces for change. The Digital Gap Initiative is one such activist group that is working 'for an inclusive digital world' and I was able to speak with Gisele Mesnage, one of the founders of the group. Gisele has a visual disability and has been campaigning in multiple ways for better technologies, including digital ones. We talked about the reasons for some ongoing challenges in digital technologies:

*Jess*: I wonder if one of the reasons it's taking some time for these corporations, and for governments to get corporations to be held to account, is that we sometimes think of the digital environment as less real than the physical geography that we're in?

*Gisele*: I think so. Also, I think we see it as moving so fast that it wouldn't be possible for the law or standards to keep up with the pace of change. Nobody can and I think this is what's so overwhelming to a lot of people. I mean, the Internet started maybe 30 years ago, and moved on to mobile phones, moved on to voice activated technology and now moving on to artificial intelligence and virtual reality. Just in the space of a couple of decades, the changes have been absolutely phenomenal. But in my view, this is probably what makes our cause realisable... I remember having a conversation with a team at Woolworths [major grocery store in Australia], and I think I can name them here, and we were sitting around the table with their IT people and their policy people and one person said, "Well, you don't understand. You give the example of access ramps in buildings and so forth...But the digital environment and its changes are so fast, there's no way we could keep that up" and I had to think quickly what do I say? Well, I said, 'look,

> actually it is the very fact that it's fast changing that allows you to introduce standards and legislation because the digital environment is so flexible. You take a building that's maybe many years old or a building that's 60 stories tall and you're trying to make an access ramp and structurally from an engineering view, that is more difficult to do than upgrading a virtual space like a website, in my view.
> There's no reason that if you can change your website overnight, you can't also make it accessible overnight. It's just coding (Gisele Mesnage interview, October 2018)

That last comment, 'It's just coding', pierces the more-than-real's perceived exceptionalism. The normalising of the framing of digital technologies peels back the smoothed-over veneer of these tools and demands better from corporations who sell them and governments that regulate them (or don't). The everydayness of Gisele's activism demonstrates the ongoing need for interventions to close the 'digital gap', or what could be reframed as digital gaps. Standards and legislation that control the built environment need to be put in place for the built *digital* environment to remove impediments that people with disabilities frequently experience.

International human rights instruments have helped identify and support the achievement of particular human rights, including those relating to people with disabilities. The United Nation's Convention on the Rights of Persons with a Disability (CPRD) includes reference to technologies in Article 9: 'To promote the design, development, production and distribution of accessible information and communications technologies and systems at an early stage, so that these technologies and systems become accessible at minimum cost' (United Nations, n.d.). Sensitive design to meet diverse needs should begin at the point of creating digital technologies and systems. Like other United Nations conventions, the CPRD lacks enforceability in Australian law but it is significant in showing accessibility within digital technologies is recognised as an international human right relevant to all signatory states. Australia has ratified the CPRD and is therefore subject to its intentions.

The overlooked aspects of everyday life that facilitate easy access and movement for many mirror some of the changes that are needed in digital spaces. In a discussion about designing technology for inclusion of people with disabilities on Future Tense, an Australian Broadcasting Corporation radio show (ABC 2017), the strengths of design that come from providing universal access was apparent:

> Jutta Treviranus: There is this phenomenon that we call the curb cut advantage or the transfer effect. Many of the things that are innovations that we rely upon and depend upon were initially designed to address the challenges that someone with a disability has. We actually within the Inclusive Design Research Centre define disability as a mismatch between the needs of the individual and the service product or environment offered. So it's not a personal trait, it simply a consequence of not designing something for someone.
>
> And so the more we make it possible for people with needs that have not been addressed or individuals that have difficulty using a particular design in the design process, the better the design will be and the more innovative the design will be. And the benefit is felt not just by the person with the disability but it is felt by the whole population because we will have designs that are more adaptive, that are far more generous in terms of the functionality.

Similar arguments about the benefit of meeting the needs of particular groups and then securing better outcomes for all have carried over into gender mainstreaming policy and theoretical spaces (McLean and Maalsen 2017).

Universal design was first pushed by disability rights advocates and legislation (Audirac 2008). Spaces created under the principles of universal design are meant to increase inclusivity regardless of multiple axes of difference, including age, ability, race, gender, and sexuality. Universally designed spaces tend to have flexibility and are able to be used by the greatest amount of people (Gossett et al. 2009). However, despite the progressive nature of universal design, issues still arise when there is a discrepancy between the digital technology and the socio-cultural and political contexts in which they must work, as outlined in Table 11.1. For example, Imrie (2012) describes the barriers to travel still encountered by a person with a disability despite accessibility technology ranging from a lack of public transport options to an altercation over rights to board between passengers due to the delay caused to board (Imrie 2012, 877). So no amount of good universal design will help if it is not accompanied by social, cultural and political transformations along multiple axes of oppression.

**Table 11.1** Dynamics of universal design in digital technologies (developed from interviews with disability activists and Elias 2011)

| Universal inclusive design principle | Digital technology example | Problem with technology | Example of change required |
|---|---|---|---|
| Equitable use | Website | Can not be read by all | Screen reader technology for visually impaired |
| Flexibility in use | Apps | Individual uses not facilitated | App makers facilitate individual's needs through adaptive design |
| Simple and intuitive use | Albert EFTPOS machine | Touchscreen impossible for visually impaired | Devices with tactile interfaces |
| Perceptible information | Video | Not all necessary information communicated | Provide captioning to provide meaningful information for the audience |
| Tolerance for error | Educational software | Processes breakdown if the user takes an alternative pathway | Background information is given to support users when problems arise |
| Low physical effort | Digital reports and slides | Too cluttered, poor colour choices exhausts the reader/listener | Simple clear design that reads easily |
| Size and space for approach and use | Websites | Buttons and links are well spaced and not too small | Responsive web design |

## Disability Activism, Digital Activism and Digital Human Rights

Universal design is a desirable quality of digital technologies but it is unfortunately not always a priority for those producing digital infrastructure. People with disabilities advocate for change in, and change of, digital spaces in a range of ways, including making direct complaints to corporations, and engaging with formal rights processes. For example, at the Human Rights and Technology Conference Blind Citizens of Australia Chief Executive Officer Emma Bennison spoke about how some recent digital technologies have a liberatory effect while others produce new problems (Nott 2018). A stark example of the two coming together arose when she was trying to use a lift in Sydney's CBD and had to call on Aira, a visual assistive technology (AT), to navigate its use as shown in Box 11.1.

**Box 11.1: Emma Bennison's narrative of using AI to navigate poorly designed digital infrastructure**

'This morning, I was staying at a hotel and I walked into the lift and, lo and behold, there was no Braille, not even any large print, on the buttons, but then again at least there were buttons. I'm not complaining; it's great that there were buttons. I did have an experience of a touchscreen lift not too long ago, so I'm not complaining.

However, I called AIRA, which is a remote visual interpretation service for people who are blind or vision impaired, and that service at the moment is provided from the US. I called them up and I said, "I'm in this lift." I had to change to "elevator" because they didn't know what I was talking about. They thought I wanted to order a lift.

So when we got that language issue sorted out, I said, "I'm in an elevator and I can't figure out where the ground button is because there's no tactile markings" and [Oh, you wear a pair of glasses and you can use your phone's camera. I had the glasses on.] And he said, "Just tip your head forward a little bit." So I tipped my head forward. "If you just move your finger up one button, you'll be on the ground button", and off I went.

She was able to guide me out of the elevator and into the reception area where I met my friend. Very exciting, totally game changing.

We talked before about the ability of technology to create an emotional response. I have to say that probably not since the time that I got my first personal computer in 1988 have I experienced such a strong emotional response to a piece of technology as I have to this, because it really does spell liberation for me' (Australian Human Rights Commission 2018).

Clearly, Bennison felt elation and excitement at being able to get sufficient support to navigate through a poorly designed space that was far from inclusive. Still, the creation of a digitally mediated lift that is not universally designed is simply bad practice. It is noteworthy that the means to overcome such a problem is an individually accessed, digitally mediated support service. Bennison should not have to encounter such everyday challenges but it is reassuring that there are digital facilities to help overcome these; the responsibility for making sure that more-than-real enabled spaces are accessible and navigable should not be pushed on to users as this deepens the 'digital disability divide' (Dobransky and Hargittai 2016, 19). The digital disability divide refers to the exclusion that people with disabilities experience with respect to using digital tools compared to others and it is deeply entrenched, largely as digital divides map on to other inequalities (Dobransky and Hargittai 2016). The reasons for digital exclusion include 'the design of technology and the pace of technological change' (Dobransky and Hargittai 2016, 19). However, disability rights activists such as those working for 'Digital Gap' challenge this claim that the pace of technological change causes exclusion as the rapid creation and dissolution of digital technologies can actually facilitate a reduction in the digital disability divide, IF designers prioritised such ends as a part of that constant remaking.

Legal processes are in place to help people with disabilities challenge digital exclusion. The Australian Human Rights Commission has a complaints process that allows advocates to bring issues with companies or government agencies to them for resolution. In 2017–2018, the Commission received 2046 complaints about discrimination and breaches of human rights, conducted 1262 conciliation processes and nearly three quarters of these were successfully mediated (AHRC 2018). Activists who are responding to the challenges that new digital technologies produce for navigating everyday lives, from entering lifts to buying food online, frequently have to pursue changes to enable their use of these technologies. Gisele, who began Digital Gap, described her experiences following a failed conciliation process to redress exclusionary technologies that had to be resolved in court:

> In particular it was that there were no required standards, there were no laws. For example, I'd be telling my lawyers, "but under the Disability Discrimination Act, under the guidelines, it says that all websites should be accessible by the end of 2014 or something like that" and they'd say "yeah, but those are guidelines. They're not legislation and it's not hard law. This is soft law." and then I would point out "But, but we've got the standards, we've got the web accessibility standards 2.0AA." Again though they would explain this is soft law, those are voluntary standards. They still come under guidelines. There is nothing to obligate, you know, Coles [major grocery store in Australia] or any other business to actually abide by all these and you can go to court and you can get compensation but you won't get a court order to fix the website according to all those standards because it's not part of hard law. And I felt, as you also talk about, we are in this digital time - where we are relying on digital technologies, we're almost forced to use digital technologies and to do banking and shopping and reading the newspaper and all these activities that before we could do in other ways, today's done in a digital way and even I think at times this is the only way.

The 'soft law' that provides guidelines for companies do not have sufficient power to make them apply the principles of universal design. If digital technologies are 'the only way' to do many mundane and necessary activities, but they are not providing opportunities for all people to use these effectively, then the 'soft laws' are doing nothing at all for ensuring equitable usage. It is a stark injustice that people have to prosecute complaints in formal processes, either with the Commission or in courts, in order to simply use digital technologies that many others are able to take for granted. The onus is on individual consumers and/or activist groups to chase companies to apply universal design principles. Accordingly, the downside of the more-than-real is visible in this instance as the always changing, never finished digital tends to reform to provide universal access following concerted appeals by activist individuals and groups.

## Using Multiple Strategies to Change the More-Than-Real

Activism by people with disabilities relies on a range of strategies to produce effective interventions with respect to digital technologies. For some people with disabilities, digital spaces make up a substantial part of their everyday lives and the difficulties they encounter in trying to use them prompts action in the form of political campaigns, the aforementioned complaints process through the AHRC, and civil suits. Some of these processes are facilitated by digital technologies while others rely on face-to-face representations. As Gisele reports:

> My world is mostly online and again, it's one of the, I guess, dangers of doing everything online and those online campaigns like petitions that are online and email campaigns, and SMS campaigns, and all that. I'm not sure if they have impact. It can be a lot faster than organizing something that is more physical. I don't know if they can be as effective as meeting somebody face-to-face like we're meeting now. I could have had this conversation with you online. Right? I don't think we can just do activism online. I think we need to have the face-to-face.

The digital is useful and yet not in any absolute sense for Gisele. Her world, as someone with a visual disability, has become 'mostly online' and that has some constraints as well as benefits. The paradoxical situation of her digital materialities (Pink et al. 2016) marks Gisele's lived experience. Deep entanglements with digital technologies do not mean that she ignores the power of face-to-face and embodied engagement.

Disability rights activism has changed to take advantage of the affordances of social media (Pearson and Trevisan 2015). For example, in the United Kingdom, disability activists were effective in getting coverage during the 2012 Paralympics in London mostly through social media (Pearson and Trevisan 2015). There are obvious tensions here if people with disabilities are having to rely on digital spaces to get leverage when, at least in Australia, the digital inclusion gap between people with disabilities and the broader population is large and actually increased in 2018 (Thomas et al. 2018). For Thomas et al.'s (2018, 9) study on digital inclusion, people with a disability were defined as those receiving 'disability support pension (DSP) from Centrelink, or the disability pension from the Department of Veterans' Affairs' (Australian government institutions). For the purposes of a statistical analysis of digital inclusion, this definition can be partially justified, although it clearly has its limitations as people with disabilities may not be able to get welfare support. Crip/queer theory might read only analysing the circumstances of those on a pension as a part of the neoliberal inclusionism that fails to see the many axes of difference that shape people's lives.

Digital technology is clearly exacerbating some barriers to facilities that reduce full social and cultural involvement of people with disabilities, including social media where much digital activism happens. In 2012, Hollier reported that all popular social media platforms—including Facebook, Twitter and YouTube—are inaccessible to some degree. Media Access Australia, the organisation that commissioned Hollier's (2012) report, provides a list of tips to support people with disabilities to use social media. These tips are practical and clear, including recommendations such as which software is best for particular social media and what support services are available for different platforms.

Thinking again about the literature on crip theory and politics, and the activist modes underpinning this thinking, it is worthwhile spending time with Hamraie and Fritsch's (2019) *Crip Technoscience Manifesto*. Doing so will provide further insights into how to rework the more-than-real. There are four main principles in Hamraie and Fritsch's (2019) Manifesto:

- In technoscience, centre 'the work of disabled people as knowers and makers' (Hamraie and Fritsch 2019, 7).
- Access is reframed as friction—rather than allowing for inclusion politics to extend neoliberal processes, crip technoscience leaves space for contestation and refusal.
- A politics of interdependence: 'technoscience can simultaneously be entangled with global networks of domination and also provide opportunities for kinship and connection' (Hamraie and Fritsch 2019, 12).
- Disability justice is at the core of Crip Technoscience.

If the access challenges that disability activists put forward are addressed properly, then more just outcomes are possible. Likewise, if the work of disabled people as knowers and makers is centred in digital technologies, it is less likely that disability injustices will emerge.

## Digital Dilemmas and People with Disabilities: Capitalist Contradictions

Most people with disabilities opt out of digital spaces if they are inaccessible rather than complain about them (Media Access Australia, 2017). The means of making a complaint are often unclear in the digital environment and there is substantial work involved in lodging complaints and following them through to completion. As Ahmed (2017) writes in considering the politics of complaints: 'You have to persist with it: a complaint requires dealing with the consequences of complaint'. The consequences of complaints against digital technologies are tied up, therefore, in a need for persistent engagement. *Chasing up* complaints is part of the process of *making* a complaint and both add layers of complexity to doing simple, everyday things like buying groceries or using social media to connect with family and friends.

The capitalist contradictions that have already been raised in this chapter are not the only issues that dictate how people with disabilities can use digital technologies. The double bind of social media being both an avenue of seeking change and containing exclusionary spaces is a strong theme in my conversations with disability activists, and in the literature. Beyond this, there are further problems. For example, the Australian National Disability Insurance Scheme (NDIS), which is managed by the National Disability Insurance Agency, has governance policies that are unclear and illogical, including demarcating what AT is and is not from the Agency's perspective. The NDIS was a welcome development in helping people with disabilities and their families manage their needs when it was introduced at the federal level in 2016 (Buckmaster 2017). The NDIS will provide assistance to 460,000 people when fully implemented and aims to help people with disabilities to: 'access mainstream services, such as health, housing and education; access community services, such as sports clubs and libraries, and; maintain informal supports, such as family and friends' (Buckmaster 2017, 1). However, the rules governing such provision of access are quite arbitrary, including determining what digital technologies are or are not considered as appropriate AT. The overarching political rhetoric of the NDIS is about empowerment and inclusion, supposedly enabling people with disabilities to have some choice and control (Goggin and Wadiwel 2014) over the way that government funding for support needs is spent rather than being subject to one size fits all decision-making.

However, the NDIS does not seem to be living up to this rhetoric, at least in some respects. An example of these rules coming into effect is how iPads are not considered appropriate AT. In a question and answer section of the NDIS website (NDIS 2017), the following exchange is shared:

> **Q. Are iPads approved in a plan in special circumstances, and if so, can we buy it first and then get reimbursed?**
>
> A. Generally, the NDIS won't fund the purchase of an iPad. If there are certain apps that relate directly to your goals and support needs, and are beneficial to reaching those goals, the cost of those apps can be funded by your NDIS plan.
>
> All approved supports need to meet the criteria of Section 34 of the NDIS Act (2013) (external), which is about Reasonable and Necessary Supports.
>
> In order to be considered reasonable and necessary, a support must:
>
> - be related to the participant's disability
> - not include day-to-day living costs that are not related to a participant's disability support needs
> - represent value for money
> - be likely to be effective and beneficial to the participant, and
> - take into account informal supports given to participants by families, carers, networks, and the community.

> An iPad can be a great educational and communicative tool for a lot of people. Some people use certain iPad applications to help them overcome barriers they face in society because of their disability. The device is not what helps people reach their NDIS goals but rather the applications specific to the disability. (NDIS 2017)

Delineating what is, and is not, appropriate AT is determined in a top-down manner by the NDIS who controls the funds available to support people with disabilities. An iPad may be a more affordable option than other AT but because of official government policy, this device is considered inappropriate. Further confusing this policy position, the NDIS will fund the use of the apps, and not the hardware to run the apps. The liberatory potential of digital technologies, and the NDIS as well, is firmly mediated by a governance decision rather than by policies that support self-determination of people with disabilities. Research on how self-determination can be supported by effective digital technologies shows that flexibility in approaches to AT (Pacheco et al. 2018) and the capacity to choose devices is important. We could also look to clear definitions of what AT is and is not from the literature on disabilities. According to Scherer (2002, 1) 'Assistive technologies are meant to help people in their primary functional tasks'. That definition does not differentiate between whether these are commercially produced devices or otherwise and gives further impetus to the claim that this decision should not be taken out of the control of people with disabilities. AT does not need to be a digital technology that only those with disabilities need to use, but the Australian government has dictated that this is appropriate in this instance.

The other tension here is that while corporations may be producing digital devices and software that addresses the needs of people with disabilities, governments may not necessarily be supporting the uptake of these in funding packages. The dissonance between what a corporation defines as providing quality AT and how the Australian government defines the same, and what people with disabilities themselves define as AT needs, culminates in a situation where people with disabilities experience frustration at mixed messages and poor support. The not-for-profit group Blind Citizens Australia has a comprehensive web page devoted to helping people navigate the NDIA and the NDIS (Blind Citizens Australia 2018), including how to get assistance when government help fails. We can see the more-than-real becoming very strange in this context as gaps are clear between what is possible and what is supported in terms of everyday lives of people with disabilities.

## Conclusion

The more-than-real provides contradictory opportunities for people with disabilities to navigate. At times liberatory and at others restrictive, digital technologies are facilitating greater accessibility while also closing down options for more conveniences if not universally designed. As the examples in this chapter have shown, the onus to correct the inaccessibility of digital technologies often falls on people with disabilities rather than regulation by government intervention, legal processes or ad hoc corporate responsiveness.

As the Digital Gap activist succinctly argued when expressing her frustration at digital geographies failing to account for differently positioned people, it's *just coding* that really needs to change, for those with relative privilege in terms of financial support and personal liberty. For others who have limited resources and freedom, much more needs to change. Even though the digital geographies that shape everyday lives are constantly changing and are made of components that can transform with relative ease, people repeatedly experience problems with the tools that create these geographies, from simple web-based transactions to digital devices to purchase items in stores. Justice issues are evident in the digital gaps that persist for people with disabilities and the lack of *just coding* is almost unbelievable given the ease with which these transformations could be made. But they are also clear in the structures that are produced, socially, culturally, politically and economically, that produce people with disabilities as not having agency, and being unable to know and make their technological futures (Hamraie and Fritsch 2019).

The Australian government continues to make disconcerting decisions about disabilities supports, as the case of the NDIS decision-making shows. Rather than supporting appropriate and feasible technologies such as iPads, the government chooses to only support the apps that this hardware runs. AT is defined in a top-down, unilateral way that defies rationality. The more-than-real abounds with these strange positions that hinder rather than help those that need access to digital technologies. This chapter only surveys a small portion of the literature on the biopolitics of neoliberal inclusionism and how people with disabilities are claiming territory in digital geographies. Crip technoscience is continuing to open up productive contestations of digital technologies that are imposed upon people with disabilities, and subverted for ignoring the lived experiences of people who know and can make their own technological solutions. Thinking of access as friction is a generative way to think though the strange positions that society, economies and cultures impose upon people with disabilities.

## References

Australian Broadcasting Corporation. (2017). Designing Technology to Increase Inclusion for the Disabled. *Future Tense*. Retrieved August 20, 2018, from https://www.abc.net.au/radionational/programs/futuretense/designing-technology-to-increase-inclusion-for-the-disabled/9100392#transcript.

Ahmed, S. (2017). Complaint as Diversity Work. *Feminist Kill Joys*. https://feministkilljoys.com/2017/11/10/complaint-as-diversity-work/.

Audirac, A. (2008). Accessing Transit as Universal Design. *Journal of Planning Literature, 23*(8), 4–16.

Australian Human Rights Commission. (2018). *Annual Report 2017–2018*. Accessed October 10, 2018, from https://www.humanrights.gov.au/sites/default/files/document/publication/AHRC_Annual_Report_2017-2018.pdf.

Australian Human Rights Commission. (2018). Inclusion Riders: People with Disabilities and Technology. *From the Human Rights and Technology Conference*. Retrieved June 10, 2019, from https://www.youtube.com/watch?v=RAGNlWosRWc.

Blind Citizens Australia, (2018). *NDIS*. Retrieved October 5, 2018, from https://www.bca.org.au/national-disability-insurance-scheme/

Buckmaster, L. (2017). *The National Disability Insurance Scheme: A Quick Guide*. Research Paper Series 2016–2017. Retrieved August 10, 2018, from http://apo.org.au/system/files/74201/apo-nid74201-74186.pdf.

Dobransky, K., & Hargittai, E. (2016). Unrealized Potential: Exploring the Digital Disability Divide. *Poetics, 58,* 18–28.

Elias, T. (2011). Universal Instructional Design Principles for Mobile Learning. *The International Review of Research in Open and Distributed Learning, 12*(2), 143–156.

Ellis, K., Goggin, G., & Kent, M. (2015). FCJ-188 Disability's Digital Frictions: Activism, Technology, and Politics. *Fibreculture Journal, 26,* 7–31.

Goggin, G., & Wadiwel, D. (2014). Australian Disability Reform and Political Participation. In *Reform and Rhetoric in Australian Social Policy Symposium*. Retrieved April 1, 2019, from http://www.australianreview.net/digest/2014/09/goggin_wadiwel.html.

Gossett, A., Gossett, A., Mirza, M., Barnds, A. K., & Feidt, D. (2009). Beyond Access: A Case Study on the Intersection Between Accessibility, Sustainability, and Universal Design. *Disability and Rehabilitation: Assistive Technology, 4*(6), 439–450.

Hamraie, A., & Fritsch, K. (2019). Crip Technoscience Manifesto. *Catalyst: Feminism, Theory, Technoscience, 5*(1), 1–34.

Hollier, S. (2012). *SociABILITY: Social Media for People with a Disability: Report for Media Access Australia*. Retrieved May 10, 2019, from https://apo.org.au/node/28618.

Imrie, R. (2012). Universalism, Universal Design and Equitable Access to the Built Environment. *Disability and Rehabilitation, 34*(10), 873–882.

Kafer, A. (2013). *Feminist, Queer, Crip*. Bloomington: Indiana University Press.

McLean, J., & Maalsen, S. (2017). 'We Don't Want It to Be Like That for Her Again': Gendered Leadership and Online Feminism in Australian Politics and Planning. *Australian Planner, 54*(1), 24–32. https://doi.org/10.1080/07293682.2017.1297316.

Media Access Australia. (2017). Excluded Web Visitors Often Don't Complain They Just Leave. *Media Access News*. Retrieved March 31, 2019, from https://mediaaccess.org.au/latest_news/news/excluded-web-visitors-often-don%E2%80%99t-complain-they-just-leave.

Mitchell, D., & Snyder, S. (2015). *The Biopolitics of Disability: Neoliberalism, Ablenationalism, and Peripheral Embodiment*. Michigan: University of Michigan Press.

National Disability Insurance Scheme. (2017). *Weekly Q and A June 2017*. Retrieved May 10, 2018, from https://www.ndis.gov.au/news/qanda-16june.html.

Nott, G. (2018, August 3). People with Disability Both Liberated and Left Behind by Today's Tech. *Computerworld*, published. Retrieved August 5, 2018, from https://www.computerworld.com.au/article/644728/people-disability-both-liberated-left-behind-by-today-tech/.

O'Neal Irwin, S. (2016). *Digital Media: Human-Technology Connection*. Lexington: Lexington Press.

Pacheco, E., Lips, M., & Yoong, P. (2018). Transition 2.0: Digital Technologies, Higher Education, and Vision Impairment. *The Internet and Higher Education, 37*, 1–10.

Pearson, C., & Trevisan, F. (2015). Disability Activism in the New Media Ecology: Campaigning Strategies in the Digital Era. *Disability & Society, 30*(6), 1–17.

Pink, S., Ardevol, E., & Lanzeni, D. (2016). *Digital Materialities: Design and Anthropology*. London: Bloomsbury Academic.

Scherer, M. (2002). The Change in Emphasis from People to Person: Introduction to the Special Issue on. *Assistive Technology, Disability and Rehabilitation, 24*(1–3), 1–4. https://doi.org/10.1080/09638280110066262.

Steele, L., & Thomas, S. (2014). Disability at the Periphery: Legal Theory, Disability and Criminal Law. *Griffith Law Review, 23*(3), 357–369. https://doi.org/10.1080/10383441.2014.1017916.

Thomas, J., Barraket, J., Wilson, C. K., Cook, K., Louie, Y. M., & Holcombe-James, I., et al. (2018). *Measuring Australia's Digital Divide: The Australian Digital Inclusion Index 2018*. RMIT University, Melbourne, for Telstra. Retrieved April 5, 2019, from https://digitalinclusionindex.org.au/wp-content/uploads/2018/08/Australian-digital-inclusion-index-2018.pdf.

United Nations. (n.d.). *Convention on the Rights of People with Disabilities*. Retrieved June 17, 2018, from https://www.un.org/development/desa/disabilities/convention-on-the-rights-of-persons-with-disabilities/article-9-accessibility.html.

# 12

# Conclusion: Thinking with the More-Than-Real

Let's return for a moment to the front cover image. The red blossom, a flower known as a bottlebrush or callistemon in Australia, is held in space and time in the centre of a smartphone frame, and it usually appears in spring and summer in this part of the world. As these plants are hardy and frost resistant they are well matched to the boom-bust climatic patterns of southeastern Australia, where droughts were once in constant cycle with periodic floods, and winter freezing temperatures can turn nightly dew to ice. Yet the changing climate is altering these biological realities. The callistemon that lined the front of the house I grew up in are still there, producing the red flowers that entice lorikeets to the yard, shielding the kitchen from view of passers-by.

The particular flower on the cover doesn't come from those trees in my family garden: my sister-in-law took this photo in late autumn with a callistemon bloom bought from a florist. That bottlebrush either travelled a long way—from northern Australia, perhaps—or was hot-housed. The assembling of this flower before it was captured by a smartphone, and photographed again, involves shadow places and unknown trajectories. Further, the layering of meaning behind this ordinary image of rural Australia is mediated by digital geographies


J. McLean, *Changing Digital Geographies*,
https://doi.org/10.1007/978-3-030-28307-0_12

and off-screen infrastructure. Whether it evokes feelings of familiarity, or a sense of being home, will depend on where you come from and currently live, and despite this, the digital geographies that produced it are well known to many. The front cover image is carefully curated to give a sense of tricky digital geographies, punctuated by a human hand, showing a representation of an environment that is more-than-real, out of season, out of context, but still aesthetically pleasing and speaking to a sense of place. The more-than-real evokes similar dissonances as the story of the cover of this book—there are gaps, loose ends, missing links (Massey 2005), unknown trajectories, power relations and sudden stark presences in messy digital geographies.

This book has shown some ways in which the more-than-real is worked with, and against, by a range of individuals and organisations. I was privileged to learn from Indigenous scholars, people with disabilities, feminists, bureaucrats, activists, environmentalists, lawyers, digital justice activists, human rights advocates, and technology innovators; several of the interview participants hold more than one of these identities. Many more people could have been interviewed too, and there were people that I approached who did not have the time or interest to engage with this research. Future research could centre queer and LGBTQIA+ perspectives in relation to the more-than-real, building on the digital geographic literature on these positionings and ways of being. Research on the experiences of migrants and refugees with respect to digital (in)justices and the more-than-real could be pursued, and the way children and older people live in, and against, more-than-real worlds would be compelling. In the style of Haraway (1998), this book is a partial perspective on digital geographies that does not attempt to be all-encompassing or universal in any way. It offers a situated engagement with the hybrid digital geographies (Foth et al. 2013) that form everyday lives and the sometimes extraordinary things that can emerge from these connections and disconnections.

The limitations of the arguments in this book also arise from my privileged positioning in the Global North. I have been fortunate to be born and educated in a relatively prosperous context, moving from my hometown of Mudgee in rural New South Wales to Sydney to study at university before the turn of the twenty-first Century. My first experience of the more-than-real was in Mudgee High School library with a trained librarian during a lunch break. The librarian explained that the computer was able to connect to a satellite and communicate with other computers around the world via the World Wide Web. I can't recall what we looked at or found out but the peculiar sensation of this moment remains with me over twenty years later. We did not have a computer at home at this stage, and technologies like wi-fi were still science fiction, the idea of anything other than illness going viral (Rose 2016) an unimagined possibility. The librarian was tense and I didn't understand how the information was being shared yet we both knew that *something*, however ordinary, was happening.

The more-than-real idea threading through this book offers a way of looking at digital geographies that grapples with complex relations between people, cultures and environments. Digital geographies—formed through human and digital technologies relations—are changing humans and non-humans. The research that shapes this book shows that emotion and affect are important aspects of how people are working to change digital geographies, as well as how digital technologies are being used to challenge environmental sustainability, politics and sociocultural relations. From the crowdfunding not-for-profit work of organisations like GetUp! and the Climate Council, to government work by the eSafety Commission, considering why people are motivated to pursue a range of digital activity is a strong focus for many seeking justice of different sorts, including environmental justice and social justice.

Mobilising groups and individuals to engage with campaigns or educational opportunities in digital technologies requires attunement to this complex terrains of emotion, affect and reason. Digital feminists draw on these interactions as a source of strength in building powerful collectives that shift and morph in different contexts. The transformation of anger and grief into action is feasible when the malleability of digital technologies becomes central to feminist strategizing against sexism and misogyny. At the same time, resisting the constraints of the structures that allow these actions is at the forefront of digital feminists' work.

As the case studies in this book illustrate, activists, researchers and institutions are already working with the contradictions of the more-than-real in productive and creative ways. Digital justice movements are seeking to correct the overreach of corporations and governments who enthusiastically take up digital technologies with all their disruptive potentials. Governments are sometimes too eager to adopt AI initiatives, such as automating decision-making, without addressing structural flaws and inbuilt biases, let alone the problems that arise with wrong decision-making via the digital (Galloway 2017; McLean and Mackenzie 2019). Groups like Digital Rights Watch and Electronic Frontiers Australia are highlighting problems with digital technologies that exploit peoples' vulnerabilities and tend to exacerbate inequalities. For instance, the Australian example of robodebt collections demonstrates how governments can fail to protect citizens from stressful and unjust accusations, as 1 in 5 of the debts were false. Digital justice advocates, in this case, are working directly with the more-than-real to overcome injustices produced by digital technologies. Such contradictions are embedded in the polar possibilities of digital spaces as they are capable of channelling productive and generative excesses, bringing and alleviating harm in diverging ways.

Digital geographies build our understanding of global environmental change by showing how Anthropocene framings are formed. The grieving for a planet that is changing as a result of human action—including calling out the problems of these universalistic framings (Demos 2017)—is found in digital spaces where people are feeling the Anthropocene. We can also see that digital spaces are affording opportunities to build alliances to combat deleterious environmental change. But these technologies continue to pressure limited resources and rely on externalising the costs of the inbuilt obsolescence of technologies. The more-than-real gives room to understand these polarising possibilities and, as a political strategy, can carry the contradictions of digital geographies.

The more-than-real concept allows for questions to be brought together in geographical research that are often held in separate dialogues, as the concept opens up connections between material and discursive aspects in multiple ways, while challenging tendencies to 'unreal' the digital. By building on Massumi's (1987) work of the more-than-real and following on from the more-than-human, this politically strategic term aims to unpack obfuscations that sideline appreciation of the environmental and social aspects of digital geographies in diverse settings. The more-than-real also begs the question of what the 'real' is, and how these digital geographies are changing this idea. Are things real if they are embodied and tactile? Is reality defined by ideas of authenticity? Being real is not about being original, following Massumi's thinking, and it isn't necessarily in opposition to fakes. These are questions that need working and thinking through further work and, perhaps, with different cases.

My research suggests that digital geographies have entangled material and discursive dimensions that are important for sustainability and equity, and that exclusions and environmental harms are reproduced with digital technologies. Yet changes in digital geographies tend to emerge as a result of action by those who are significantly disadvantaged by new technologies, rather than corporations or governments, at this point in time and space. Disability activists have argued that digital technologies, similarly to other technologies, are designed and implemented in a way that is often inherently exclusionary and that much more needs to be done to achieve equity in digital spaces. The 'digital gap' is substantial but the affordances of digital technologies are also considerable; again, the more-than-real's paradoxes are clear in this area of digital (in)justice. The work of correcting such inequities should not lie with activists and not-for-profits: corporations and governments need to ensure that the coding underlying digital innovations is accessible by design, and that their exercise of biopower does not reduce accessibility. It is clearly unjust and ridiculous to expect that those with disabilities should bear the burden of highlighting inaccessibility and challenging, corporations and governments to remedy these on a case-by-case basis.

I return to Eve Kosofsky Sedgwick's ideas about theories that are beside rather than beyond, to suggest that I have offered the more-than-real as an additional concept to think with, and on, the messiness of digital spaces. Sedgwick (2002, 8) says that she wants to invoke 'Deleuzian interest in planar relations, the irreducibly spatial positionality of beside also seems to offer some useful resistance to the ease with which beneath and beyond turn from spatial descriptors into implicit narratives of, respectively, origin and telos'. The spatiality of

Sedgwick's appeal to planar relations connects with this work on the more-than-real. Digital technologies produce spaces that are complex intertwinings of humans and more-than-humans, technologies and ephemera, emotions and affect. The literature on digital geographies argues that there is a digital turn underway, an inflection point that requires attention and explanation. I add to this conceptualisation of a shift in geographic thinking to suggest caution on the use of a 'turn' as it suggests a core of geographic thought and an external periphery that now seeks prominence. The language of 'turns' also has a directionality that is suggestive of linear thinking while digital geographic scholarship is, or should be, always placed in specific locations and hence diverse.

Digital justice stories are emerging from work on AI and surveillance cases and more can be done in this area. Decentring of humans from digital rights is a possibility if we take heed of the narratives and reflections collated in this book, from activists, academics, organisers and bureaucrats. Among others, feminists and legal activists who are reworking digital geographies are conscious of environmental harms and are keen to work towards more sustainable technologies. At the same time, the non-humans of digital geographies are frequently overlooked and green IT is largely a sidelined and/or tokenistic concern for corporations and governments. The digital Anthropocene we find ourselves in is not a place we can comfortably call home, keeping in mind the troubling qualities of any imagined category such as 'we'.

Indigenous activists, academics and writers are already remaking digital geographies, challenging colonial hegemonies with creative, constructive and persistent challenges to the status quo. As Chapter 5 shows, connections to Country, assertions of Indigenous presence and excellence, and Indigitalising software and hardware, are all core to these efforts at producing digital geographies that are inclusive and diverse. While threats of neocolonial extension do persist in digital geographies, counter-challenges to colonial hegemonies are pervasive and growing too.

## What Now? Keep Reworking the More-Than-Real

The research participants for this book offered pragmatic and achievable suggestions and reflections on the strengths and weaknesses of digital geographies that are worthwhile bringing together here. First, bringing digital rights conversations together with formal legal frameworks was a repeated theme. Rather than leaving digital corporations to self-regulate, or indeed to not regulate at all, governments need to increase their control over digital geographies to protect citizens. On the other hand, governments should not be uncritically adopting disruptive technologies through AI and other means, or impinging on the privacy of individuals by collecting metadata and creating laws that allow unfettered access by digital corporations, and perpetuating digital injustices themselves. More than once it was suggested that digital justice needs to be embedded in a Bill of Rights approach in Australia, and globally digital justice conversations must continue to grow.

Decolonising digital geographies may not be a desirable or feasible proposition but bringing to the fore Indigenous knowledges, engagement and representations was a strong theme of this research. Destabilising settler colonial practices in digital technology was definitely a common concern in the literature on Indigenous technologies and digital engagement (for example, Carlson and Frazer 2018). Working to resist, refuse and overturn colonial hegemonies is everyone's work, Indigenous or not, and the permeability of digital geographies can help this process. Returning to Latour (2014), and thinking of how we could apply his challenge to nurture digital technologies rather than neglect them, part of loving our digital monsters involves regulatory processes to rework our capitalist systems to make better digital spaces, with good enough data (Daly et al. 2019).

Pulling apart corporate claims of sustainability for digital technologies is an important and ongoing project that needs more consideration, along with current concerns about expansion of AI and big data in contexts including government overreach and corporate control. As the more-than-real reaches its slippery tentacles even deeper into everyday lives, data consumption rates around the world show no signs of slowing, while the singular Resource Man continues to emerge as a utopic solution to growing energy use (Strengers 2013). Holding corporations accountable for poor environmental responsibility in the more-than-real needs to become the work of governments. While we are expanding debates around human rights into digital technology realms, the sidelining of sustainability concerns in these conversations is concerning. It is exciting to see conversations about human rights and technology growing, such as those led by government, corporate and not-for-profit global institutions, where humanising the digital is the main priority. But the shape of these conversations can also morph to make sure we avoid the same mistakes of other human rights dialogues: the more-than-human must be a part of these dialogues and the actions they hope to achieve, if we seek to move away from anthropocentric decision-making. The discourse of sustainability, as human-centred as it is, does not make a notable appearance in important digital technology and human rights conversations, such as that being led by the Australian Human Rights Commission.

I think that the language that is used to think on, through, and with, the digital can keep changing, too. Terms like 'virtual reality' reinforce the not-realness of digital geographies, as does the wry use of acronyms like IRL. Immersive technology is a better term for virtual reality as it touches on the entanglement of seemingly being there, but not, giving us a way to talk about the heightened emotional and embodied responses that those devices can give. Perhaps the ethical dilemmas that are associated with digitisation may be better navigated with terms that avoid casting the digital as unreal. There is nothing particularly new about these suggestions—as I show in Chapter 2, much has already been said about the binaries we fall into when discussing digital technologies and concepts like the Anthropocene.

Donna Haraway was recently profiled in *The Guardian* and she told stories of her academic path, including a telling exchange at a conference in Brazil that she attended with Latour:

> It was a bunch of primate field biologists, plus me and Bruno. And Stephen Glickman, a really cool biologist, took us apart privately. He said: "Now, I don't want to embarrass you. But do you believe in reality?"

> We were both kind of shocked by the question. First, we were shocked that it was a question of belief, which is a Protestant question. A confessional question. The idea that reality is a question of belief is a barely secularized legacy of the religious wars. In fact, reality is a matter of worlding and inhabiting. It is a matter of testing the holdingness of things. Do things hold or not? (Weigel 2019, paragraphs 11–12)

The question of whether things hold or not, and how reality is made by worlding and inhabiting, resonates with the stories told in this book. The worlding that is happening in and with digital geographies, that is made possible by digital technologies, sits in relation to older debates about the binaries between nature and humans, the Global South and the Global North. Meanwhile, the efforts to make just worlds, where multi-kinds, as Haraway calls them, might coexist harmoniously, rather than in conflict, is also frequently emerging from digital spaces. Wrangling the more-than-real to be an agent for that generative work forms part of those efforts.

## References


Carlson, B., & Frazer, R. (2018). Yarning Circles and Social Media Activism. *Media International Australia, 169*(1), 43–53. https://doi.org/10.1177/1329878X18803762.

Daly, A., Devitt, K., & Mann, M. (2019). *Good Data*. Brisbane/Meanjin: Queensland University of Technology. Retrieved June 19, 2019, from http://networkcultures.org/wp-content/uploads/2019/01/Good_Data.pdf.

Demos, T. J. (2017). *Against the Anthropocene: Visual Culture and Environment Today*. Berlin: Sternberg Press.

Foth, M., Parra Agudelo, L., & Palleis, R. (2013). Digital Soapboxes: Towards an Interaction Design Agenda for Situated Civic Innovation. In *Proceedings of the 2013 ACM Conference on Pervasive and Ubiquitous Computing Adjunct Publication* (pp. 725–728). ACM. Accessed June 10, 2019, from https://www.uni-siegen.de/infme/start_ifm/veranstaltungen/marcus_foth_-_position_paper.pdf.

Galloway, K. (2017). Big Data: A Case Study of Disruption and Government Power. *Alternative Law Journal, 42*(2), 89–95. https://doi.org/10.1177/1037969X17710612.

Haraway, D. (1998). Situated Knowledges: The Science Question in Feminism and the Privilege of Partial Perspective. *Feminist Studies, 14*(3), 575–599.

Latour, B. (2014). Love Your Monsters. *Next Nature*. Accessed December 4, 2018, from http://www.nextnature.net/2014/09/love-your-monsters/.

Massey, D. (2005). *For Space*. London: Sage.

Massumi, B. (1987). Realer Than Real: The Simulacrum According to Deleuze and Guattari. *Copyright, 1,* 90–97. http://www.brianmassumi.com/textes/REALER%20THAN%20REAL.pdf.

McLean, J., & Mackenzie, R. (2019). Digital Justice in Australian Visa Application Processes? *Alternative Law Journal*. https://doi.org/10.1177/1037969X19853685.

Rose, G. (2016). Cultural Geography Going Viral. *Social & Cultural Geography, 17*(6), 763–767. https://doi.org/10.1080/14649365.2015.1124913.

Sedgwick, E. K. (2002). *Touching Feeling: Affect, Pedagogy, Performativity*. Durham: Duke University Press.

Strengers, Y. (2013). *Smart Energy Technologies in Everyday Life: Smart Utopia?* London: Palgrave Macmillan.

Weigel, M. (2019). Feminist Cyborg Scholar Donna Haraway: 'The Disorder of Our Era Isn't Necessary'. *The Guardian*. Retrieved June 22, 2019, from https://www.theguardian.com/world/2019/jun/20/donna-haraway-interview-cyborg-manifesto-post-truth.

# Appendix

This book draws on 15 interviews with activists, startup founders, digital safety designers, lawyers and academics. Some research participants listed below by name chose to waive their anonymity after reviewing content for publication. Many thanks to all research participants for their generous time and fascinating reflections on their work.

| Type of work/organisation | Role | Interview date |
|---|---|---|
| Digital feminist activism | Moderator | June 2018 |
| Digital startup, Hello Cass | Founder/Director | September 2018 |
| eSafety Commissioner | Julie Inman Grant, Commissioner | September 2018 |
| Digital safety monitoring | Support officer | September 2018 |
| Climate change action NGO | Community Engagement officer | October 2018 |
| Communicate climate science, Climate Council | Lesley Hughes, Councillor | November 2018 |
| Environmental NGO | Founder | October 2018 |
| Cyber security, University | Researcher | October 2018 |
| Disabilities rights activist/ digital gap | Gisele Mesnage, Founder and activist | October 2018 |
| Privacy protection/Australian privacy foundation | Mark Waldom, Secretary of APF | October 2018 |



J. McLean, *Changing Digital Geographies*,
https://doi.org/10.1007/978-3-030-28307-0

| Type of work/organisation | Role | Interview date |
|---|---|---|
| Promoting and protecting digital rights, Electronic Frontiers Australia | Lyndsey Jackson, Chair of EFA | October 2018 |
| Digital technologies researcher, Macquarie University | Prof. Deb Richards | October 2018 |
| Indigenous knowledges and digital technologies, Macquarie University | Prof. Bronwyn Carlson | November 2018 |
| Digital rights activist, Digital Rights Watch Australia | Lizzie O'Shea, lawyer and board member | November 2018 |
| Independent, issue-based organisation, Get Up! Board member | Lyn Goldsworthy, board member GetUp! | November 2018 |

# Index




J. McLean, *Changing Digital Geographies*,
https://doi.org/10.1007/978-3-030-28307-0

The manufacturer's authorised representative in the EU is Springer Nature Customer Service Centre GmbH, Europaplatz 3, 69115 Heidelberg, Germany. If you have any concerns regarding our products, please contact ProductSafety@springernature.com

Printed and bound by CPI Group (UK) Ltd, Croydon, CR0 4YY
25/09/2026
02222859-0003